Uncertainty Quantification Techniques in Statistics

Uncertainty Quantification Techniques in Statistics

Special Issue Editor

Jong-Min Kim

MDPI • Basel • Beijing • Wuhan • Barcelona • Belgrade • Manchester • Tokyo • Cluj • Tianjin

Special Issue Editor
Jong-Min Kim
University of Minnesota at Morris
USA

Editorial Office
MDPI
St. Alban-Anlage 66
4052 Basel, Switzerland

This is a reprint of articles from the Special Issue published online in the open access journal *Mathematics* (ISSN 2227-7390) (available at: https://www.mdpi.com/journal/mathematics/special_issues/uncertainty_quantification_techniques_statistics).

For citation purposes, cite each article independently as indicated on the article page online and as indicated below:

LastName, A.A.; LastName, B.B.; LastName, C.C. Article Title. *Journal Name* **Year**, *Article Number*, *Page Range*.

ISBN 978-3-03928-546-4 (Pbk)
ISBN 978-3-03928-547-1 (PDF)

© 2020 by the authors. Articles in this book are Open Access and distributed under the Creative Commons Attribution (CC BY) license, which allows users to download, copy and build upon published articles, as long as the author and publisher are properly credited, which ensures maximum dissemination and a wider impact of our publications.

The book as a whole is distributed by MDPI under the terms and conditions of the Creative Commons license CC BY-NC-ND.

Contents

About the Special Issue Editor . . . . **vii**

Preface to "Uncertainty Quantification Techniques in Statistics" . . . . **ix**

Javier E. Contreras-Reyes, Mohsen Maleki and Daniel Devia Cortés
Skew-Reflected-Gompertz Information Quantifiers with Application to Sea Surface Temperature Records
Reprinted from: *Mathematics* **2019**, *7*, 403, doi:10.3390/math7050403 . . . . **1**

Md Showaib Rahman Sarker , Michael Pokojovy and and Sangjin Kim
On the Performance of Variable Selection and Classification via Rank-Based Classifier
Reprinted from: *Mathematics* **2019**, *7*, 457, doi:10.3390/math7050457 . . . . **15**

Sangjin Kim and Jong-Min Kim
Two-Stage Classification with SIS Using a New Filter Ranking Method in High Throughput Data
Reprinted from: *Mathematics* **2019**, *7*, 493, doi:10.3390/math7060493 . . . . **31**

Gi-Sung Lee, Ki-Hak Hong and Chang-Kyoon Son
An Estimation of Sensitive Attribute Applying Geometric Distribution under Probability Proportional to Size Sampling
Reprinted from: *Mathematics* **2019**, *7*, 1102, doi:10.3390/math7111102 . . . . **47**

Abhijeet R Patil and Sangjin Kim
Combination of Ensembles of Regularized Regression Models with Resampling-Based Lasso Feature Selection in High Dimensional Data
Reprinted from: *Mathematics* **2020**, *8*, 110, doi:10.3390/math8010110 . . . . **63**

Jung Yeon Lee, Myeong-Kyu Kim, Wonkuk Kim
Robust Linear Trend Test for Low-Coverage Next-Generation Sequence Data Controlling for Covariates
Reprinted from: *Mathematics* **2020**, *8*, 217, doi:10.3390/math8020217 . . . . **87**

Hohsuk Noh and Seong J. Yang
Comparing Groups of Decision-Making Units in Efficiency Based on Semiparametric Regression
Reprinted from: *Mathematics* **2020**, *8*, 233, doi:10.3390/math8020233 . . . . **101**

About the Special Issue Editor

Jong-Min Kim (Dr.) is currently working as a Full Professor in the Statistics Division at the Science and Mathematics University of Minnesota-Morris, Minnesota, USA. He received his Ph.D. (Statistics) in 2002 from the Department of Statistics, Oklahoma State University, Stillwater, Oklahoma, USA (Minor: Mathematics). He worked as a Research Fellow at SAMSI-The Statistical and Applied Mathematical Sciences Institute (NSF, Duke, NCSU, and UNC). He received the Morris Faculty Distinguished Research Award. His research involves statistical genetics, cluster analysis, big data analytics, text mining for patent data, copula directional dependence, and cryptocurrency data analysis. He has published about 120 refereed journal research papers in statistics, biostatistics, bioinformatics, and economics. He is Guest Editor, Associate Editor, and Editorial Board member of several peer reviewed journals. Some of his important research publications in the field of copula directional dependence and big data analytics are listed at: https://academics.morris.umn.edu/jong-min-kim.

Preface to "Uncertainty Quantification Techniques in Statistics"

Uncertainty quantification (UQ) is a mainstream research topic in applied mathematics and statistics. To identify UQ problems, diverse modern techniques for large and complex data analyses have been developed in applied mathematics, computer science, and statistics. This Special Issue of Mathematics (ISSN 2227-7390) includes diverse modern data analysis methods such as skew-reflected-Gompertz information quantifiers with application to sea surface temperature records, the performance of variable selection and classification via a rank-based classifier, two-stage classification with SIS using a new filter ranking method in high throughput data, an estimation of sensitive attribute applying geometric distribution under probability proportional to size sampling, combination of ensembles of regularized regression models with resampling-based lasso feature selection in high dimensional data, robust linear trend test for low-coverage next-generation sequence data controlling for covariates, and comparing groups of decision-making units in efficiency based on semiparametric regression.

Jong-Min Kim
Special Issue Editor

Article

Skew-Reflected-Gompertz Information Quantifiers with Application to Sea Surface Temperature Records

Javier E. Contreras-Reyes [1,*], Mohsen Maleki [2] and Daniel Devia Cortés [3]

1 Departamento de Estadística, Facultad de Ciencias, Universidad del Bío-Bío, Concepción 4081112, Chile
2 Department of Statistics, College of Sciences, Shiraz University, Shiraz 71946 85115, Iran; m.maleki.stat@gmail.com
3 Departamento de Evaluación de Pesquerías, Instituto de Fomento Pesquero, Valparaíso 2361827, Chile; ddeviac@gmail.com
* Correspondence: jcontreras@ubiobio.cl or jecontrr@uc.cl; Tel.: +56-41-311-1199

Received: 19 March 2019; Accepted: 2 May 2019; Published: 6 May 2019

Abstract: The Skew-Reflected-Gompertz (SRG) distribution, introduced by Hosseinzadeh et al. (J. Comput. Appl. Math. (2019) 349, 132–141), produces two-piece asymmetric behavior of the Gompertz (GZ) distribution, which extends the positive to a whole dominion by an extra parameter. The SRG distribution also permits a better fit than its well-known classical competitors, namely the skew-normal and epsilon-skew-normal distributions, for data with a high presence of skewness. In this paper, we study information quantifiers such as Shannon and Rényi entropies, and Kullback–Leibler divergence in terms of exact expressions of GZ information measures. We find the asymptotic test useful to compare two SRG-distributed samples. Finally, as a real-world data example, we apply these results to South Pacific sea surface temperature records.

Keywords: Skew-Reflected-Gompertz distribution; Gompertz distribution; entropy; Kullback–Leibler divergence; sea surface temperature

---

1. Introduction

The Skew-Reflected-Gompertz (SRG) distribution was recently introduced by [1] and corresponds to an extension of the Gompertz distribution [2], named after Benjamin Gompertz (1779–1865). It extends the positive dominion $\mathbb{R}_+$ to the whole of $\mathbb{R}$ by an extra parameter, ε, $-1 < \varepsilon < 1$, and produces two-piece asymmetric behavior of Gompertz (GZ) density. The SRG distribution has as particular cases the Reflected-GZ and GZ distributions, when $\varepsilon \to 1$ and $\varepsilon \to -1$, respectively. The SRG distribution family can also represent a suitable competitor against the skew-normal (SN, [3]) and epsilon-skew-normal (ESN, [4]) distributions as a way to fit asymmetrical datasets. Indeed, refs. [5,6] dealt with the frequentist and Bayesian inferences of ESN distribution. Contributions by [1] provided probability density function (pdf), cumulative distribution function (cdf), quantile function, moment-generating function (MGF), stochastic representation, the Expectation-Maximization (EM) algorithm for SRG parameter estimates and the Fisher information matrix (FIM).

Moreover, several recent investigations confirmed the usefulness of entropic quantifiers in the study of asymmetric distributions [3,7,8] and their applications to topics such as thermal wake [9], marine fish biology [3,8], sea surface temperature (SST), relative humidity measured in the Atlantic Ocean [10], and more. We build on the study of [3], which developed hypothesis testing for normality, i.e., if the shape parameter is close to zero. They considered the Kullback–Leibler (KL) divergence in terms of moments and cumulants of the modified SN distribution. Posteriorly, we consider a real-world data set of the anchovy condition factor for testing the shape parameter to decide if a food deficit produced by environmental conditions such as El Niño exists [11].

This work arose from a motivation to tackle the problem of determining the adequate pdf of SST [9,10]. Indeed, probabilistic modelling of SST is key for accurate predictions [9]. Therefore, we propose that the SRG model based on two-piece distributions could be more suitable for interpreting annual bimodal and asymmetric SST data. We also considered the existent results of Shannon and Rényi entropies, and KL divergence for GZ distributions for developed entropic quantifiers for SRG distributions. Posteriorly, we considered SST along the South Pacific and Chilean coasts from 2012 to 2014 to illustrate our results. Specifically, we introduced hypothesis testing developed by [12] for the SRG distribution, which is useful to compare two data sets with bimodal and asymmetric behavior such as SST.

2. The Skew-Reflected-Gompertz Distribution

The Gompertz (GZ, [2]) distribution is a continuous probability distribution with the following pdf

$$f(x|\sigma,\eta) = \frac{\eta}{\sigma} e^{\frac{x}{\sigma}} e^{-\eta(e^{\frac{x}{\sigma}}-1)}, \quad x \geq 0, \tag{1}$$

where $\sigma > 0$ and $\eta > 0$ are the scale and shape parameters, respectively, and are denoted by $X \sim GZ(\sigma,\eta)$. The mean and variance of X are

$$\begin{aligned} E(X) &= \sigma e^{\eta} Ei(-\eta), \\ Var(X) &= \sigma^2 e^{\eta} \tau, \end{aligned} \tag{2}$$

respectively; where $Ei(z) = \int_{-z}^{\infty} \frac{e^{-u}}{u} du$, $\tau = -2\eta F(-\eta) + \gamma^2 + \frac{\pi^2}{6} + 2\gamma \log \eta + (\log \eta)^2 - e^{\eta}[Ei(-\eta)]^2$, $\gamma = 0.5772156649$ is the Euler constant and

$$F(z) = \sum_{k=0}^{+\infty} \frac{z^k}{k!(k+1)^3}.$$

The SRG distribution is an extension of the GZ proposed by [1]. If Y follows, the SRG distribution is denoted by $Y \sim SRG(\mu,\sigma,\eta,\varepsilon)$ and has pdf

$$g(y|\mu,\sigma,\eta,\varepsilon) = \begin{cases} \frac{1}{2} f\left(\frac{\mu-y}{1+\varepsilon}\Big|\sigma,\eta\right), & y \leq \mu, \\ \frac{1}{2} f\left(\frac{y-\mu}{1-\varepsilon}\Big|\sigma,\eta\right), & y > \mu, \end{cases} \tag{3}$$

where $\mu \in \mathbb{R}$ is the location parameter and $\varepsilon \in (-1,1)$ is the slant parameter. Note that SRG is the GZ distribution when $\mu = 0$ and $\varepsilon \to -1$, GZ distribution with negative support when $\varepsilon \to 1$, and Reflected-GZ distribution when $\varepsilon = 0$. Also, the Reflected-GZ distribution corresponds to a particular case of a more general class of two-piece asymmetric distributions proposed by [13,14]. The mean, variance and MGF of Y are

$$\begin{aligned} E(Y) &= \mu - 2\varepsilon\sigma e^{\eta} Ei(-\eta), \\ Var(Y) &= \sigma^2 \{\tau e^{\eta} + 2(1-\varepsilon^2)e^{2\eta}[Ei(-\eta)]^2\}, \\ M_Y(t) &= \frac{1}{2}\eta e^{\eta+\mu t}[(1-\varepsilon)F_{-\sigma t}(\eta) + (1+\varepsilon)F_{\sigma t}(\eta)], \end{aligned} \tag{4}$$

respectively; where $F_s(z) = \int_1^{\infty} v^{s+1} e^{-vz} dv$. Jafari et al. [15] provide the MGF of X using expansion series. However, (4) is considered a clearer expression that depends only on integral $F_s(z)$. See Section 4.1 for some details of the MLE EM-based algorithm related to SRG parameters.

According to [1], the SRG distribution can be re-parametrized in terms of GZ and Reflected-GZ distributions as

$$g(y|\mu,\sigma_+,\sigma_-,\eta) = p_1 f(\mu - y|\sigma_+,\eta)I_{(-\infty,\mu]}(y) + p_2 f(y - \mu|\sigma_-,\eta)I_{(\mu,+\infty)}(y), \tag{5}$$

where $\sigma_\pm = \sigma(1\pm\varepsilon)$, $p_1 + p_2 = 1$, and $p_1 = \sigma_+/(\sigma_+ + \sigma_-) = (1+\varepsilon)/2$. Let $\mathbf{Y} = (Y_1,\ldots,Y_n)^\top$ be an i.i.d sample from the SRG distribution with parameters $(\mu, \sigma_\pm, \eta)$ and latent vectors $\mathbf{Z} = (\mathbf{Z}_1,\ldots,\mathbf{Z}_n)$, thus (5) can be equivalently represented as $(-1)^j (Y_i - \mu)|Z_{ij} = 1 \sim GZ(\sigma_\pm, \eta)$, $i = 1,\ldots,n, j = 1,2$, where $\mathbf{Z}_i = (Z_{i1}, Z_{i2})^\top \sim \text{Mult}(1, p_1, p_2)$ is a multinomial vector, $P(Z_{i1} = z_{i1}, Z_{i2} = z_{i2}) = p_1^{z_{i1}} p_2^{z_{i2}}$, $z_{ij} = \{0,1\}$, and $z_{i1} + z_{i2} = 1$. Given that $P(Z_{i1} = 1) = P(Z_{i1} = 1, Z_{ik} = 0; \forall j \neq k)$, the complete log-likelihood function is

$$\ell(\mu, \sigma_+, \sigma_-, \eta | \mathbf{Y}, \mathbf{Z}) = -n\log(2\sigma) + n(\eta + \log \eta)$$
$$+ \sum_{i=1}^{n} \left[z_{i1}\left(\frac{\mu - y_i}{\sigma_+} - \eta e^{\frac{\mu - y_i}{\sigma_+}}\right) + z_{i2}\left(\frac{y_i - \mu}{\sigma_-} - \eta e^{\frac{y_i - \mu}{\sigma_-}}\right) \right]. \quad (6)$$

Conditional expectations of latent variables $\mathbf{Z}_i$ are given by

$$\widehat{z}_{i1} = E[Z_{i1}|\widehat{\mu}, \widehat{\sigma}_+, \widehat{\sigma}_-, y_i] = \widehat{p}_1 \frac{f(\widehat{\mu} - y_i|\widehat{\sigma}_+, \widehat{\eta})}{g(y_i|\widehat{\mu}, \widehat{\sigma}_+, \widehat{\sigma}_-, \widehat{\eta})} I_{(-\infty,\widehat{\mu}]}(y_i), \quad (7)$$

$$\widehat{z}_{i2} = 1 - \widehat{z}_{i1}, \quad i = 1,\ldots,n. \quad (8)$$

The E- and M-steps on the $(k+1)$th iteration of the EM algorithm are

E-step. From (6)–(8), we have

$$Q(\mu, \sigma_+, \sigma_-, \eta | \mu^{(k)}, \sigma_+^{(k)}, \sigma_-^{(k)}, \eta^{(k)}) = E[\ell(\mu, \sigma_+, \sigma_-, \eta | \mathbf{Y}, \mathbf{Z}) | \mu^{(k)}, \sigma_+^{(k)}, \sigma_-^{(k)}, \eta^{(k)}]$$
$$= -n\log(2\sigma) + n(\eta + \log \eta)$$
$$+ \sum_{i=1}^{n} \left[\widehat{z}_{i1}^{(k)}\left(\frac{\mu - y_i}{\sigma_+} - \eta e^{\frac{\mu - y_i}{\sigma_+}}\right) + \widehat{z}_{i2}^{(k)}\left(\frac{y_i - \mu}{\sigma_-} - \eta e^{\frac{y_i - \mu}{\sigma_-}}\right) \right].$$

and

M-step. Update $\sigma_\pm$, by solving the following equation

$$\sum_{i=1}^{n} \widehat{z}_{ij}^{(k)} \left(\eta^{(k)} \frac{|y_i - \mu^{(k)}|}{\sigma_\pm^2} e^{\frac{|y_i - \mu^{(k)}|}{\sigma_\pm}} - \frac{|y_i - \mu^{(k)}|}{\sigma_\pm^2} \right) = \frac{n}{2\sigma}.$$

Update μ by solving the following equation

$$\widehat{\mu}^{(k+1)} = \operatorname*{argmax}_{\mu} \sum_{i=1}^{n} \left\{ \widehat{z}_{i1}^{(k)}\left(\frac{\mu - y_i}{\widehat{\sigma}_+^{(k+1)}} - \eta e^{\frac{\mu - y_i}{\widehat{\sigma}_+^{(k+1)}}}\right) + \widehat{z}_{i2}^{(k)}\left(\frac{\mu - y_i}{\widehat{\sigma}_-^{(k+1)}} - \eta e^{\frac{\mu - y_i}{\widehat{\sigma}_-^{(k+1)}}}\right) \right\}.$$

Update η by

$$\widehat{\eta} = n \left(\sum_{i=1}^{n} \left\{ \widehat{z}_{i1}^{(k)} e^{\frac{\mu - y_i}{\widehat{\sigma}_+^{(k+1)}}} + \widehat{z}_{i2}^{(k)} e^{\frac{\mu - y_i}{\widehat{\sigma}_-^{(k+1)}}} \right\} \right)^{-1}.$$

The EM-algorithm must be iterated until the sufficient convergence rule is satisfied:

$$\|(\widehat{\mu}^{(k+1)}, \widehat{\sigma}_+^{(k+1)}, \widehat{\sigma}_-^{(k+1)}, \widehat{\eta}^{(k+1)}) - (\widehat{\mu}^{(k)}, \widehat{\sigma}_+^{(k)}, \widehat{\sigma}_-^{(k)}, \widehat{\eta}^{(k)})\| < \tau,$$

for a tolerance τ close to zero. The FIM for standard deviations of MLEs $(\widehat{\mu}, \widehat{\sigma}, \widehat{\eta}, \widehat{\varepsilon})$ and additional details of the EM-algorithm are described in [1].

3. Entropic Quantifiers

In the next section, we present the main results of entropic quantifiers for SRG distribution.

3.1. Shannon Entropy

The Shannon entropy (SE), introduced by [16] in the context of univariate continuous distributions, quantifies the information contained in a random variable X with pdf $f(x)$ through the expression

$$H(X) = -\int_{-\infty}^{+\infty} f(x) \log f(x) dx. \tag{9}$$

The SE concept is attributed to the uncertainty of the information presented in X [17]. Propositions 1 and 2 present the SE for GZ and SRG distributions, respectively.

Proposition 1. *[15]. The SE of $X \sim GZ(\sigma, \eta)$ is*

$$H(X) = \log\left\{\frac{B(1,1)}{\eta}\right\} - \sigma\eta - \frac{E(X)}{\sigma} + \sigma\eta M_X(\sigma^{-1}),$$

where $B(\cdot, \cdot)$ is the usual Beta function and $E(X)$ is given in (2).

Substituting $\mu = 0$ and $\varepsilon = -1$ into (4) (i.e., reducing SRG to its special case GZ), we obtain $M_X(\sigma^{-1}) = \eta e^{\eta} F_{-1}(\eta) = 1$. Therefore, $H(X)$ in Proposition 1 is reduced to

$$H(X) = -\log \eta - e^{\eta} E_i(-\eta), \tag{10}$$

i.e., the SE of the GZ random variable only depends on shape parameter η.

Proposition 2. *The SE of $Y \sim SRG(\mu, \sigma, \eta, \varepsilon)$ is*

$$H(Y) = \frac{1+\varepsilon}{2}\left\{H(X_{+\varepsilon}) - \log\left(\frac{1+\varepsilon}{2}\right)\right\} + \frac{1-\varepsilon}{2}\left\{H(X_{-\varepsilon}) - \log\left(\frac{1-\varepsilon}{2}\right)\right\},$$

where $X_{\pm\varepsilon} \sim GZ(\sigma(1 \pm \varepsilon), \eta)$ and $H(X_{\pm\varepsilon})$ are obtained using Proposition 1.

Proof. From (3) and (9), we obtained

$$\begin{aligned}
H(Y) &= -\int_{-\infty}^{+\infty} g(y|\mu, \sigma, \eta, \varepsilon) \log g(y|\mu, \sigma, \eta, \varepsilon) dy \\
&= -\frac{1}{2}\int_0^{+\infty} f\left(\frac{x}{1+\varepsilon}\bigg|\sigma, \eta\right) \log\left\{\frac{1}{2}f\left(\frac{x}{1+\varepsilon}\bigg|\sigma, \eta\right)\right\} dx \\
&\quad -\frac{1}{2}\int_0^{+\infty} f\left(\frac{x}{1-\varepsilon}\bigg|\sigma, \eta\right) \log\left\{\frac{1}{2}f\left(\frac{x}{1-\varepsilon}\bigg|\sigma, \eta\right)\right\} dx \\
&= -\frac{1}{2}\int_0^{+\infty} (1+\varepsilon) f(x|\sigma(1+\varepsilon), \eta) \log\left\{\frac{1+\varepsilon}{2}f(x|\sigma(1+\varepsilon), \eta)\right\} dx \\
&\quad -\frac{1}{2}\int_0^{+\infty} (1-\varepsilon) f(x|\sigma(1-\varepsilon), \eta) \log\left\{\frac{1-\varepsilon}{2}f(x|\sigma(1-\varepsilon), \eta)\right\} dx,
\end{aligned}$$

which concludes the proof. □

From (10), given that $H(X_{\pm\varepsilon})$ only depends on shape parameter η, we obtain $H(X_{\pm\varepsilon}) = H(X)$, and $H(Y)$ only depends on η and ε parameters. Therefore,

$$H(Y) = -\log \eta - e^{\eta} E_i(-\eta) - \frac{1+\varepsilon}{2}\log\left(\frac{1+\varepsilon}{2}\right) - \frac{1-\varepsilon}{2}\log\left(\frac{1-\varepsilon}{2}\right). \tag{11}$$

Figure 1 illustrates SE behavior for random variable Y. We observed that SE increases when η decreases. For each η, SE is maximized and minimized at $\varepsilon = 0$ (Reflected-GZ) and $\varepsilon \to -1$

(Truncated-GZ and GZ), respectively. More details appear in [3,8] for the SE expressions of other asymmetric distributions.

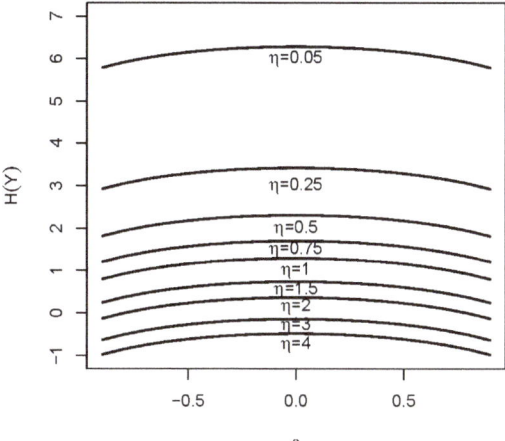

Figure 1. Shannon entropy of Skew-Reflected-Gompertz (SRG) distributions for $\varepsilon \in (-1,1)$ and several values of η.

3.2. Rényi Entropy

The αth-order Rényi entropy (RE), introduced by [18] in the context of univariate continuous distributions, extends the concept of SE information contained in a random variable X with pdf $f(x)$ through a level α, $\alpha \in \mathbb{N}$, $\alpha > 0$, and the expression

$$R_\alpha(X) = \frac{1}{1-\alpha} \log \int_{-\infty}^{+\infty} [f(x)]^\alpha dx. \tag{12}$$

RE information can be negative and is ordered with respect to α, i.e., $R_{\alpha_1}(X) \geq R_{\alpha_2}(X)$ for any $\alpha_1 < \alpha_2$ (see, e.g., [7] and other properties of RE). From (12), the SE is obtained by the limit of $H(X) = \lim_{\alpha \to 1} R_\alpha(X)$ by applying l'Hôpital's rule to $R_\alpha(X)$ with respect to α (see e.g., [7]). The RE of the GZ and SRG distributions is presented in Propositions 3 and 4, respectively.

Proposition 3. *[15,19]. The RE of $X \sim GZ(\sigma, \eta)$ with $\alpha > 1$, $\alpha \in \mathbb{N}$, is*

$$R_\alpha(X) = -\frac{\log \alpha}{1-\alpha} + \log \frac{\eta}{\sigma} + \frac{1}{1-\alpha} \log \left\{ \sum_{j=0}^{\alpha-1} \binom{\alpha-1}{j} \frac{\Gamma(j+1)}{(\alpha \eta)^j} \right\},$$

where $\Gamma(u) = \int_0^\infty t^{u-1} e^{-t} dt$ is the gamma function.

Proposition 4. *The RE of $Y \sim SRG(\eta, \varepsilon)$ with $\alpha > 1$, $\alpha \in \mathbb{N}$, is*

$$R_\alpha(Y) = \frac{1}{1-\alpha} \log \left\{ \left(\frac{1+\varepsilon}{2}\right)^\alpha e^{(1-\alpha)R_\alpha(X_{+\varepsilon})} + \left(\frac{1-\varepsilon}{2}\right)^\alpha e^{(1-\alpha)R_\alpha(X_{-\varepsilon})} \right\},$$

where $X_{\pm\varepsilon} \sim GZ(\sigma(1\pm\varepsilon), \eta)$ and $R_\alpha(X_{\pm\varepsilon})$ are obtained using Proposition 3.

Proof. From (3) and (12), we obtained

$$R_\alpha(Y) = \frac{1}{1-\alpha} \log \int_{-\infty}^{+\infty} [g(y|\mu,\sigma,\eta,\varepsilon)]^\alpha dy,$$

$$= \frac{1}{1-\alpha} \log \left\{ \int_0^{+\infty} \left[\frac{1}{2} f\left(\frac{x}{1+\varepsilon}\Big|\sigma,\eta\right)\right]^\alpha dx + \int_0^{+\infty} \left[\frac{1}{2} f\left(\frac{x}{1-\varepsilon}\Big|\sigma,\eta\right)\right]^\alpha dx \right\},$$

$$= \frac{1}{1-\alpha} \log \left\{ \left(\frac{1+\varepsilon}{2}\right)^\alpha \int_0^{+\infty} [f(x|\sigma(1+\varepsilon),\eta)]^\alpha dx + \left(\frac{1-\varepsilon}{2}\right)^\alpha \int_0^{+\infty} [f(x|\sigma(1-\varepsilon),\eta)]^\alpha dx \right\},$$

which concludes the proof. □

Figure 2a illustrates the behavior of RE for random variable Y when $\alpha = 2$ (quadratic RE). As in the SE case, we also observed that RE increases when η decreases and reaches maximum and minimum at $\varepsilon = 0$ (Reflected-GZ) and $\varepsilon \to -1$ (Truncated-GZ and GZ), respectively. When $\alpha = 5$ (or $\alpha > 2$) (see Figure 2b), RE decays faster than in the quadratic RE case as $\varepsilon \to -1$. More details appear in [7] for the RE expressions of other asymmetric distributions.

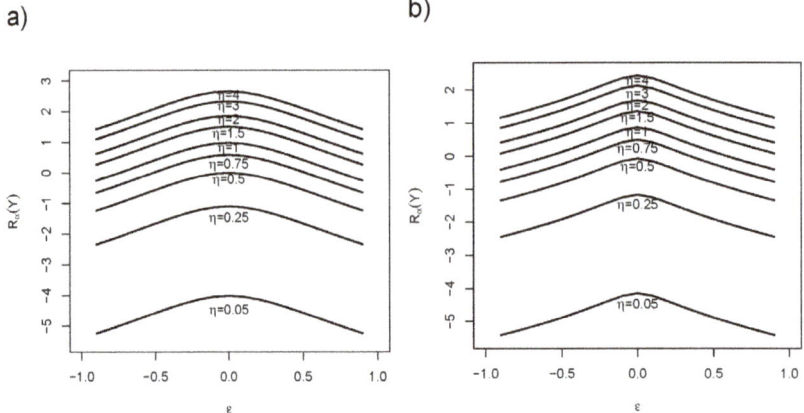

Figure 2. Rényi entropy of SRG distributions for $\sigma = 1$, $-1 < \varepsilon < 1$, several values of η and (**a**) $\alpha = 2$ and (**b**) $\alpha = 5$ values.

3.3. Kullback–Leibler Divergence

The Kullback–Leibler (KL) divergence introduced by [20] in the context of univariate continuous distributions, extends the concept of SE between two random variables X_1 and X_2 with pdfs $f_1(x_1)$ and $f_2(x_2)$, respectively, through the expression

$$K(X_1, X_2) = \int_{-\infty}^{+\infty} f_1(x) \log \left\{\frac{f_1(x)}{f_2(x)}\right\} dx. \qquad (13)$$

The KL divergence measures the disparity between the pdfs of X_1 and X_2, and is non-negative, non-symmetric and zero only if $X_1 = X_2$ in distribution. Also, the KL divergence does not satisfy the triangular inequality (see, e.g., [8,17] for other properties of KL and other divergences). The KL divergence for two GZ and two SRG distributions are presented in Propositions 5 and 6.

Proposition 5. [21]. *The KL divergence between $X_1 \sim GZ(\sigma_1, \eta_1)$ and $X_2 \sim GZ(\sigma_2, \eta_2)$ is*

$$K(X_1, X_2) = \log\left\{\frac{e^{\eta_1}\sigma_2\eta_1}{e^{\eta_2}\sigma_1\eta_2}\right\} + e^{\eta_1}\left[\left(\frac{\sigma_1}{\sigma_2} - 1\right) E_i(-\eta_1) + \frac{\eta_2}{\eta_1^{\sigma_1/\sigma_2}} \Gamma\left(\frac{\sigma_1}{\sigma_2} - 1, \eta_1\right)\right] - (\eta_1 + 1),$$

where $\Gamma(u,v) = \int_v^\infty t^{u-1} e^{-t} dt$ is the upper incomplete gamma function.

Proposition 6. *The KL divergence between $Y_1 \sim SRG(0, \sigma_1, \eta_1, \varepsilon_1)$ and $Y_2 \sim SRG(0, \sigma_2, \eta_2, \varepsilon_2)$ is*

$$K(Y_1, Y_2) = \frac{1+\varepsilon_1}{2}\left[\log\left\{\frac{1+\varepsilon_1}{1+\varepsilon_2}\right\} + K(X_{+\varepsilon_1}, X_{+\varepsilon_2})\right] + \frac{1-\varepsilon_1}{2}\left[\log\left\{\frac{1-\varepsilon_1}{1-\varepsilon_2}\right\} + K(X_{-\varepsilon_1}, X_{-\varepsilon_2})\right],$$

where $X_{\pm\varepsilon_i} \sim GZ(\sigma_i(1\pm\varepsilon_i), \eta_i)$, $i = 1, 2$, and $K(X_{\pm\varepsilon_1}, X_{\pm\varepsilon_2})$ are obtained using Proposition 5.

Proof. From (3) and (13), we obtained

$$K(Y_1, Y_2) = \int_{-\infty}^{+\infty} g(x|0, \sigma_1, \eta_1, \varepsilon_1) \log\left\{\frac{g(x|0, \sigma_1, \eta_1, \varepsilon_1)}{g(x|0, \sigma_2, \eta_2, \varepsilon_2)}\right\} dx,$$

$$= \frac{1}{2}\int_0^{+\infty} f\left(\frac{x}{1+\varepsilon_1}\Big|\sigma_1, \eta_1\right) \log\left\{\frac{f\left(\frac{x}{1+\varepsilon_1}\Big|\sigma_1, \eta_1\right)}{f\left(\frac{x}{1+\varepsilon_2}\Big|\sigma_2, \eta_2\right)}\right\} dx$$

$$+ \frac{1}{2}\int_0^{+\infty} f\left(\frac{x}{1-\varepsilon_1}\Big|\sigma_1, \eta_1\right) \log\left\{\frac{f\left(\frac{x}{1-\varepsilon_1}\Big|\sigma_1, \eta_1\right)}{f\left(\frac{x}{1-\varepsilon_2}\Big|\sigma_2, \eta_2\right)}\right\} dx,$$

$$= \frac{1+\varepsilon_1}{2}\left[\log\left\{\frac{1+\varepsilon_1}{1+\varepsilon_2}\right\} + \int_0^{+\infty} f(x|\sigma_1(1+\varepsilon_1), \eta_1) \log\left\{\frac{f(x|\sigma_1(1+\varepsilon_1), \eta_1)}{f(x|\sigma_2(1+\varepsilon_2), \eta_2)}\right\} dx\right]$$

$$+ \frac{1-\varepsilon_1}{2}\left[\log\left\{\frac{1-\varepsilon_1}{1-\varepsilon_2}\right\} + \int_0^{+\infty} f(x|\sigma_1(1-\varepsilon_1), \eta_1) \log\left\{\frac{f(x|\sigma_1(1-\varepsilon_1), \eta_1)}{f(x|\sigma_2(1-\varepsilon_2), \eta_2)}\right\} dx\right],$$

which concludes the proof. □

More details appear in [3,8] for the KL divergence expressions of other asymmetric distributions. Using Proposition 6, the asymptotic KL divergence between $Y \sim SRG(0, \sigma, \eta, \varepsilon)$ and $X \sim GZ(\sigma, \eta)$ is

$$K(Y, X) \approx \frac{1+\varepsilon}{2}\left[\lim_{\varepsilon_2 \to -1} \log\left(\frac{1+\varepsilon}{1+\varepsilon_2}\right) + K(X_{+\varepsilon}, X)\right] + \frac{1-\varepsilon}{2}\left[\log\left(\frac{1-\varepsilon}{2}\right) + K(X_{-\varepsilon}, X)\right],$$

as $\varepsilon_2 \to -1$. However, we see that $\log\left(\frac{1+\varepsilon}{1+\varepsilon_2}\right) = +\infty$ as $\varepsilon_2 \to -1$ and $K(Y, X)$ is not finite. However, from Proposition 6 the asymptotic KL divergence between Y_1 and Y_2 is

$$K(Y_1, Y_2) \approx K(X, Y) = \log\left(\frac{2}{1-\varepsilon}\right) + K(X, X_{-\varepsilon}), \quad (14)$$

as $\varepsilon_1 \to -1$, where $X_{-\varepsilon} \sim GZ(\sigma(1-\varepsilon), \eta)$. Therefore, while $K(Y, X)$ is not finite, $K(X, Y)$ is finite and can be used to study the disparity of ε from -1. Thus, hypothesis testing for $H_0: \varepsilon = -1$ can be addressed. Besides, we further study hypothesis testing for scale and shape parameters between two SRG distributions in Section 3.4. From (14), we also took that $K(Y_1, Y_2) \approx K(X, X_1)$ as $\varepsilon \to -1$, with $X_1 \sim GZ(2\sigma, \eta)$.

Figure 3 illustrates the KL divergence between two SRG distributions. We observed that for the critical points of $(\varepsilon_1, \varepsilon_2) \to \{(-1, 1); (1, -1)\}$, the KL divergence reaches the highest values and is close to zero in the other values [panels (a) and (b)]. For large η's [panel (c)], the KL divergence is zero for a concentrated region of the dominion where $\varepsilon_1 = \varepsilon_2$.

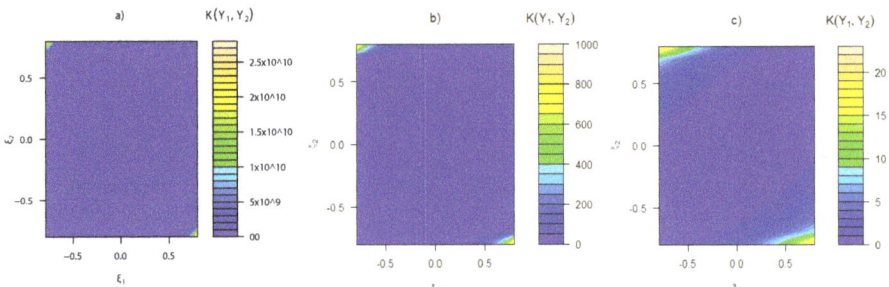

Figure 3. Plots of Kullback–Leibler (KL) divergence between $Y_1 \sim \text{SRG}(0, \sigma_1, \eta_1, \varepsilon_1)$ and $Y_2 \sim \text{SRG}(0, \sigma_2, \eta_2, \varepsilon_2)$ for values $\sigma_1 = \sigma_2 = 1$ and (a) $\eta_1 = \eta_2 = 0.25$; (b) $\eta_1 = \eta_2 = 3$; and (c) $\eta_1 = \eta_2 = 10$.

All information quantifiers and the EM algorithm for SRG distribution were implemented in [22].

3.4. Asymptotic Test

Consider two independent samples of sizes n_1 and n_2 from Y_1 and Y_2, respectively; where $\theta, \theta' \in \Theta \subset \mathbb{R}^p$, and X_1 and X_2 have pdfs $g(y; \theta_1)$ and $g(y; \theta_2)$, respectively; with $\theta_i = (\sigma_i, \eta_i, \varepsilon_i)$, $i = 1, 2$. Suppose partition $\theta_i = (\theta_{i1}, \theta_{i2})$, and assume $\theta_{21} = \theta_{11} \in \Theta_1 \subset \mathbb{R}^r$, so that $\theta_{i2} \in \Theta \cap \Theta_1^c \subset \mathbb{R}^{p-r}$. Let $\widehat{\theta}_i = (\widehat{\theta}_{i1}, \widehat{\theta}_{i2})$ be the MLE of $\theta_i = (\theta_{i1}, \theta_{i2})$ for $i = 1, 2$, which corresponds to the MLE of the full model parameters (θ_1, θ_2) under the null hypothesis $H_0 : \theta_{21} = \theta_{11}$. Thus, part b) of Corollary 1 in [12] establishes that if the null hypothesis $H_0 : \theta_{22} = \theta_{12}$ holds and $\frac{n_1}{n_1 + n_2} \xrightarrow[n_1, n_2 \to \infty]{} \lambda$, with $0 < \lambda < 1$, then

$$K_0 = \frac{2n_1 n_2}{n_1 + n_2} K(\widehat{\theta}_1, \widehat{\theta}_2) \xrightarrow[n_1, n_2 \to \infty]{d} \chi^2_{p-r}, \quad (15)$$

where $r = 3$ is the number of parameters of the SRG distribution (location parameter is not considered for KL divergence). Thus, a test of level α for the above homogeneity null hypothesis consists of rejecting H_0 if $K_0 > \chi^2_{p-r, 1-\alpha}$, where $\chi^2_{p-r, \alpha}$ is the αth percentile of the χ^2_{p-r}-distribution.

As [3] stated, the proposed asymptotic test is only valid for regular conditions of the SRG distribution, in particular for a non-singular FIM. Therefore, given that the SRG distributions' FIM is singular at $\varepsilon \to \pm 1$ [1], the SRG model does not serve for testing the null hypothesis using (15) when ε is close to -1 or 1.

4. Application

4.1. Sea Surface Temperature Data

The spatial information and SST data analyzed in this study were recorded by a scientific observer (whose labor concerns biological sampling of fishes, incidental captures of birds, turtles and marine mammals. Biological sampling was complemented with information such as time, longline and hook features, number of buoys, baits, etc.) (SO) in the Chilean longline fleet (industrial and artisanal), which was oriented to capture swordfish (*Xiphias gladius*, [23]) from 2012 to 2014 (obtaining a sampling of 83% in 2012, 55% in 2013, 90% in 2014, and 75% in 2012–2014). The covered area of the study was at 21°31'–36°39' LS and 71°08'–85°52' LW (see Figure 4).

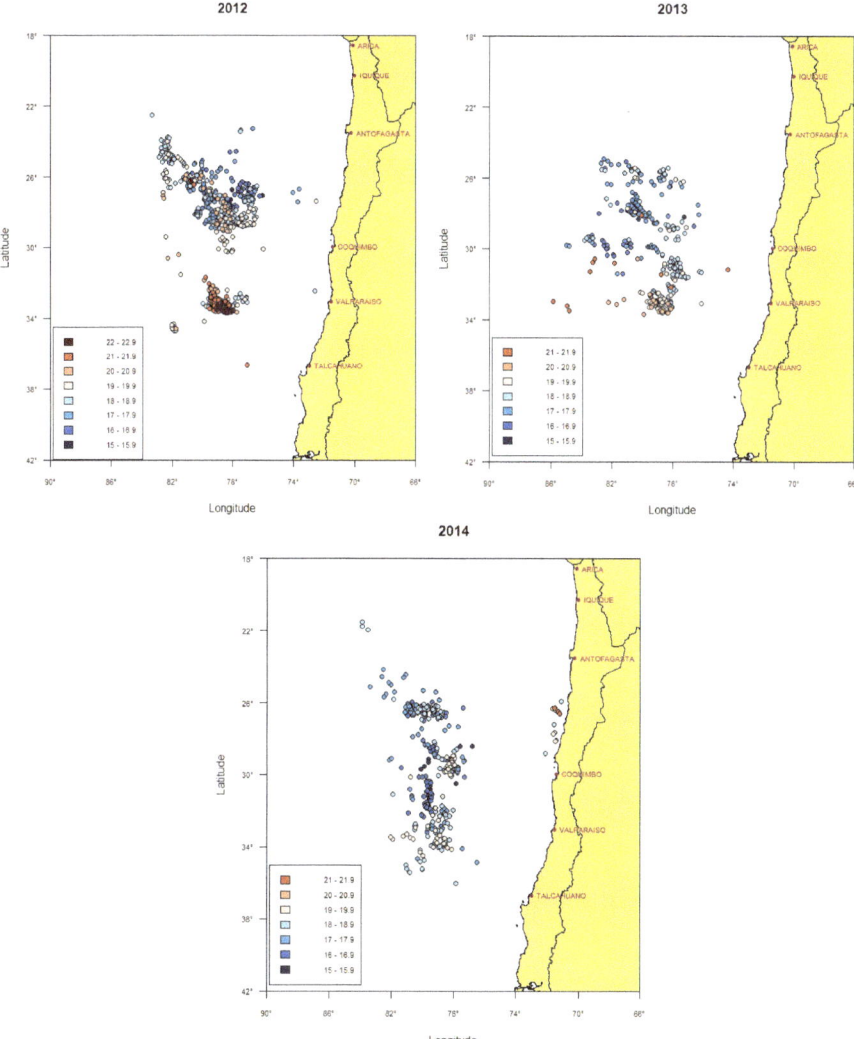

Figure 4. Spatial distribution of Sea Surface Temperature (SST) observations by year (21°31′–36°39′ LS, 71°08′–85°52′ LW).

SST records in swordfish captures are crucial for distributional analysis and fish abundance. Specifically, variations in SST are physical factors that control productivity, growth and migration of species [24]. In addition, SST is strongly correlated with atmospheric pressure at sea level and thus climatic time scales. Therefore, changes in SST overlap with ecosystem changes [25]. However, SST influence on ecosystems is not clear because other physical processes such as superficial warming, horizontal advection of currents, upwelling, etc. [11], modify SST. Therefore, SST anomalies could be symptomatic rather than causal.

4.1.1. SRG Parameter Estimates

Considering the smallest Akaike (AIC) and Schwarz (BIC) information criteria, we observed in Table 1 that SRG performs better than the SN and ESN models (see Appendices A and B, respectively). In addition, Table 1 shows the estimated parameters (based on the EM algorithm presented in Section 2) for SST datasets by year assuming SRG distribution. In 2012, a negative ε estimate corresponds to asymmetry to the right, and in 2013 and 2014 negative ε and η close to zero produce a two-piece distribution to fit "cold" and "warm" temperatures (Figure 5).

Table 1. Parameter estimates and their respective standard deviations (SD) for SST by year based on SRG, epsilon-skew-normal (ESN) and skew-normal (SN) models. For each model, log-likelihood function $\ell(\theta)$, $\theta = (\mu, \sigma, \eta, \varepsilon)$, Akaike's (AIC) and Bayesian (BIC) information criteria, and goodness-of-fit tests (Kolmogorov–Smirnov (K–S), Anderson–Darling (A–D), and Cramer–von Mises, (C–V)) are also reported with respective p-values in parentheses.

Year	Model	Param.	Estim.	(S.D)	$\ell(\theta)$	AIC	BIC	K–S	A–D	C–V
2012 ($n=774$)	SRG	μ	17.992	0.103	-1401.896	2811.793	2830.399	0.044 (0.095)	2.014 (0.090)	0.214 (0.242)
		σ	2.590	0.067						
		η	1.444	0.027						
		ε	-0.207	0.075						
	ESN	θ	18.000	0.031	-1507.534	3021.069	3035.023	0.118 (<0.01)	26.417 (<0.01)	2.059 (<0.01)
		ω	1.657	0.015						
		ϵ	-0.418	0.069						
	SN	ξ	16.777	0.114	-1404.581	2815.161	2829.116	0.041 (0.143)	1.752 (0.126)	0.198 (0.271)
		ω	5.199	0.043						
		λ	2.527	0.311						
2013 ($n=415$)	SRG	μ	17.935	0.061	-687.420	1382.839	1398.942	0.082 (0.010)	2.632 (0.042)	0.491 (0.041)
		σ	1.112	0.026						
		η	0.432	0.021						
		ε	-0.108	0.029						
	ESN	θ	17.600	0.046	-716.375	1438.750	1450.827	0.089 (<0.01)	7.721 (<0.01)	0.970 (0.002)
		ω	1.328	0.026						
		ϵ	-0.376	0.092						
	SN	ξ	16.598	0.200	-691.531	1389.063	1401.140	0.066 (0.054)	2.002 (0.092)	0.328 (0.113)
		ω	3.812	0.054						
		λ	2.421	0.617						
2014 ($n=439$)	SRG	μ	17.454	0.048	-653.082	1314.164	1330.502	0.092 (<0.01)	2.848 (0.033)	0.533 (0.032)
		σ	0.896	0.020						
		η	0.375	0.020						
		ε	-0.106	0.025						
	ESN	θ	17.200	0.053	-703.748	1413.496	1425.750	0.109 (<0.01)	11.996 (<0.01)	1.529 (<0.01)
		ω	0.956	0.035						
		ϵ	-0.384	0.090						
	SN	ξ	16.146	0.098	-666.984	1339.968	1352.222	0.096 (<0.01)	4.055 (<0.01)	0.711 (0.011)
		ω	3.245	0.045						
		λ	3.434	0.618						

To evaluate the goodness-of-fit test, the Kolmogorov–Smirnov (K–S), Anderson–Darling (A–D), and Cramer–von Mises (C–V) tests were considered for all models, commonly used to analyze the goodness-of-fit test of a particular distribution see, e.g., [26]). Considering a 95% confidence level, SRG fits perform well for 2012 and 2013, and on a 90% confidence level, the SRG fit performs well for 2014.

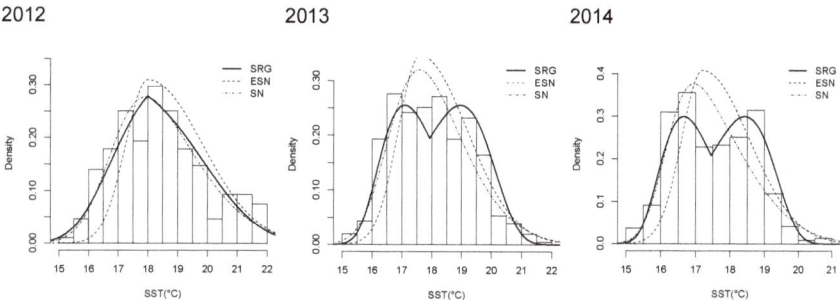

Figure 5. MLE fit of SRG, ESN and SN models for SST data by year.

4.1.2. Information Quantifiers and Asymptotic Test

Parameters estimated from the SRG model and presented in Table 1 are used to perform the quantifiers of Sections 3.1–3.3 for SST in each year and for the asymptotic test of Section 3.4 for comparing SST between two years. The results of these analyses are shown in Table 2. In Table 2, $K_0 = \widehat{K}(Y_1, Y_2)$ represents the KL divergence between the years Y_1 (column) and Y_2 (row).

The first quantifiers (SE and RE) illustrate that the highest information of SST is obtained by SE and increases with the increment of years. For all RE, the highest information of SST is obtained in 2012 and is negative for 2013 and 2014 and similar during that period. Differences in information between SE and RE are produced by the independency of SE with parameter σ, while RE depends on three parameters as in Proposition 4.

In addition, the asymptotic test presented in Table 2 is analogous for all the years in both groups. In fact, the null hypothesis $H_0 : \theta_1 = \theta_2$ is rejected at a 95% confidence level. This rejection is reinforced by high values of statistics K_0, produced by a high sample size of both groups (n_1 and n_2).

Table 2. SRG Shannon, $H(Y)$, and Rényi, $R_\alpha(Y)$, $\alpha = 2, 3, 4$, entropies for SST data. For each year, the KL divergence $K_0 = \widehat{K}(Y_1, Y_2)$, statistic and its respective p-values of Equation (15) are reported. All reported K_0 estimates considered the estimated parameters and sample size n in Table 1.

Year	Quantifier	2012	2013	2014
	$H(Y)$	0.765	0.781	2.754
	$R_2(Y)$	0.384	−0.362	−0.365
	$R_3(Y)$	0.252	−0.417	−0.418
	$R_4(Y)$	0.163	−0.457	−0.457
2012	K_0	-	0.266	0.911
	Statistic	-	143.740	520.41
	p-value	-	<0.01	<0.01
2013	K_0	0.080	-	0.071
	Statistic	43.192	-	30.233
	p-value	<0.01	-	<0.01
2014	K_0	0.143	0.043	-
	Statistic	80.327	18.282	-
	p-value	<0.01	<0.01	-

5. Conclusions

We have presented a methodology to compute the Shannon and the Rényi entropy and the Kullback–Leibler divergence for the family of Skew-Reflected-Gompertz distributions. Our methods consider the information quantifiers previously computed for the Gompertz distribution. Explicit formulas for Shannon and Rényi entropies (in terms of the Gompertz, Shannon and

Rényi entropies, respectively), and the Kullback–Leibler divergence (using incomplete gamma function) facilitate easy computational implementation. Additionally, given the regularity conditions accomplished by the Skew-Reflected-Gompertz distribution, specifically by the Fisher information matrix convergence when ε is in $(-1,1)$, an asymptotic test for comparing two groups of datasets was developed.

The statistical application to South Pacific sea surface temperature was given. We first carried out SRG goodness-of-fit tests in samples over three years, where we find strong evidence (a 95% confidence level) for 2012, and moderate evidence (a 90% confidence level) for 2013 and 2014. The results show that the proposed methodology serves to compare two sets of samples, Skew-Reflected-Gompertz distributed. The proposed asymptotic test is therefore useful to detect anomalies in sea surface temperature, linked to extreme events influenced by environmental conditions [11,24,25]. We encourage researchers to consider the proposed methodology for further investigations related to environmental datasets [1].

Author Contributions: J.E.C.-R. and M.M. wrote the paper and contributed reagents/analysis/materials tools; J.E.C.-R. and D.D.C. conceived, designed and performed the experiments and analyzed the data. All authors have read and approved the final manuscript.

Funding: This research received no external funding.

Acknowledgments: We are grateful to the Instituto de Fomento Pesquero (IFOP) for providing access to the data used in this work. Special thanks to Fernando Espíndola for his helpful insights and discussion on an early version of this paper. The SST datasets and R codes used in this work are available upon request to the corresponding author. The authors thank the editor and two anonymous referees for their helpful comments and suggestions.

Conflicts of Interest: The authors declare that there is no conflict of interest in the publication of this paper.

Abbreviations

The following abbreviations are used in this manuscript:

A–D	Anderson–Darling
AIC	Akaike's information criterion
BIC	Bayesian information criterion
C–V	Cramer–von Mises
CDF	Cumulative distribution function
EM	Expectation maximization
ESN	Epsilon-skew-normal
FIM	Fisher information matrix
GZ	Gompertz
K–S	Kolmogorov–Smirnov
KL	Kullback–Leibler
MGF	Moment-generating function
MLE	Maximum Likelihood Estimator
PDF	Probability density function
RE	Rényi entropy
SD	Standard deviation
SE	Shannon entropy
SN	Skew-normal
SRG	Skew-Reflected-Gompertz
SST	Sea surface temperature

Appendix A. The Epsilon-Skew-Normal Distribution

The epsilon-skew-normal distribution [4,27] in its location-scale version is denoted as $\text{ESN}(\theta, \omega, \varepsilon)$. It can be derived from a more general class of two-piece asymmetric distributions proposed by [14], by considering the standardized normal kernel $\phi(\cdot)$ (zero mean and variance 1), denoted as $N(0,1)$, as the density f and the functions $a(\epsilon) = 1 + \epsilon$ and $b(\epsilon) = 1 - \epsilon$. If $Z \sim \text{ESN}(\theta, \varpi, \epsilon)$, thus Z has pdf given by

$$h(z|\theta, \varpi, \epsilon) = \begin{cases} \phi\left(\frac{\theta - z}{\varpi(1+\epsilon)}\right), & z \leq \theta, \\ \phi\left(\frac{z - \theta}{\varpi(1-\epsilon)}\right), & z > \theta, \end{cases} \quad (A1)$$

where $Z = \theta + \varpi X$ for location $\theta \in \mathbb{R}$ and scale $\varpi > 0$ parameters. The mean and variance of Z are

$$E(Z) = \theta - 4\varpi\epsilon/\sqrt{2\pi},$$
$$\text{Var}(Z) = \frac{\varpi^2}{\pi}[(3\pi - 8)\epsilon^2 + \pi],$$

and the MGF of X is given by

$$M_X(t) = (1 + \epsilon)e^{\frac{(1+\epsilon)^2 t^2}{2}} \Phi[-(1+\epsilon)t] + (1 - \epsilon)e^{\frac{(1-\epsilon)^2 t^2}{2}} \Phi[(1-\epsilon)t],$$

where $\Phi(\cdot)$ is the cdf of standardized Gaussian distribution.

Appendix B. The Skew-Normal Distribution

Let X be a skew-normal (SN, [28]) random variable denoted as $X \sim \text{SN}(\xi, \omega, \lambda)$. The pdf of X is given by

$$f(x; \lambda) = 2\phi(z)\Phi(\lambda z), \quad (A2)$$

with $z = (x - \xi)/\omega$. The SN model with the density (A2) is explained by its stochastic representation

$$X \stackrel{d}{=} \xi + \delta|U_0| + \sqrt{1 - \delta^2} U, \quad (A3)$$

where $\delta = \lambda/\sqrt{1 + \lambda^2}$, X is represented as a linear combination of Gaussian U and a half-Gaussian $|U_0|$ variable, and $U_0 \sim N(0,1)$ and $U \sim N(0, \omega^2)$ are independent (Theorem 1 of [29]). From (A3), the mean and variance of X are $E(X) = \xi + \sqrt{2/\pi}\delta$ and $\text{Var}(X) = \omega^2 - (2/\pi)\delta^2$, respectively.

References

1. Hoseinzadeh, A.; Maleki, M.; Khodadadi, Z.; Contreras-Reyes, J.E. The Skew-Reflected-Gompertz distribution for analyzing symmetric and asymmetric data. *J. Comput. Appl. Math.* **2019**, *349*, 132–141. [CrossRef]
2. Gompertz, B. On the nature of the function expressive of the law of human mortality, and on a new mode of determining the value of life contingencies. *Philos. Trans. R. Soc. Lond.* **1825**, *115*, 513–583. [CrossRef]
3. Arellano-Valle, R.B.; Contreras-Reyes, J.E.; Stehlík, M. Generalized skew-normal negentropy and its application to fish condition factor time series. *Entropy* **2017**, *19*, 528. [CrossRef]
4. Mudholkar, G.S.; Hutson, A.D. The epsilon-skew-normal distribution for analyzing near-normal data. *J. Stat. Plan. Inference* **2000**, *83*, 291–309. [CrossRef]
5. Maleki, M.; Mahmoudi, M.R. Two-Piece Location-Scale Distributions based on Scale Mixtures of Normal family. *Commun. Stat. Theor. Meth.* **2017**, *46*, 12356–12369. [CrossRef]
6. Moravveji, B.; Khodadai, Z.; Maleki, M. A Bayesian Analysis of Two-Piece distributions based on the Scale Mixtures of Normal Family. *Iran. J. Sci. Technol. Trans. A* **2019**, *43*, 991–1001. [CrossRef]
7. Contreras-Reyes, J.E. Rényi entropy and complexity measure for skew-gaussian distributions and related families. *Physica A* **2015**, *433*, 84–91. [CrossRef]
8. Contreras-Reyes, J.E. Analyzing fish condition factor index through skew-gaussian information theory quantifiers. *Fluct. Noise Lett.* **2016**, *15*, 1650013. [CrossRef]

9. Wang, Y.Q.; Derksen, R.W. The confirmation of the α–β model and the maximum entropy formulation in a thermal wake. *Environmetrics* **1998**, *9*, 269–282. [CrossRef]
10. De Queiroz, M.M.; Silva, R.W.; Loschi, R.H. Shannon entropy and Kullback–Leibler divergence in multivariate log fundamental skew-normal and related distributions. *Can. J. Stat.* **2016**, *44*, 219–237. [CrossRef]
11. Di Lorenzo, E.; Combes, V.; Keister, J.E.; Strub, P.T.; Thomas, A.C.; Franks, P.J.; Ohman, M.D.; Furtado, J.C.; Bracco, A.; Bograd, S.J.; et al. Synthesis of Pacific Ocean climate and ecosystem dynamics. *Oceanography* **2013**, *26*, 68–81. [CrossRef]
12. Salicrú, M.; Menéndez, M.L.; Pardo, L.; Morales, D. On the applications of divergence type measures in testing statistical hypothesis. *J. Multivar. Anal.* **1994**, *51*, 372–391. [CrossRef]
13. Maleki, M.; Contreras-Reyes, J.E.; Mahmoudi, M.R. Robust Mixture Modeling Based on Two-Piece Scale Mixtures of Normal Family. *Axioms* **2019**, *8*, 38. [CrossRef]
14. Arellano-Valle, R.B.; Gómez, H.W.; Quintana, F.A. Statistical inference for a general class of asymmetric distributions. *J. Stat. Plan. Inference* **2005**, *128*, 427–443. [CrossRef]
15. Jafari, A.A.; Tahmaoobi, S.; Alizadeh, M. The beta Gompertz distribution. *Rev. Colomb. Estad.* **2014**, *37*, 141–158. [CrossRef]
16. Shannon, C.E. A mathematical theory of communication. *Bell Syst. Tech. J.* **1948**, *27*, 379–423. [CrossRef]
17. Cover, T.M.; Thomas, J.A. *Elements of Information Theory*; Wiley & Son, Inc.: New York, NY, USA, 2006.
18. Rényi, A. *Probability Theory*; Dover Publications: New York, NY, USA, 2012.
19. Abu-Zinadah, H.H.; Aloufi, A.S. Some characterizations of the exponentiated Gompertz distribution. *Int. Math. Forum* **2014**, *9*, 1427–1439. [CrossRef]
20. Kullback, S.; Leibler, R.A. On information and sufficiency. *Ann. Math. Stat.* **1951**, *22*, 79–86. [CrossRef]
21. Bauckhage, C. Characterizations and Kullback–Leibler Divergence of Gompertz Distributions. *arXiv* **2014**, arXiv:1402.3193.
22. R Core Team. *A Language and Environment for Statistical Computing*; R Foundation for Statistical Computing: Vienna, Austria, 2018; ISBN 3-900051-07-0.
23. Barría, P.; González, A.; Cortés, D.D.; Mora, S.; Miranda, H.; Cerna, F.; Cid, L.; Ortega, J.C. *Seguimiento Pesquerías Recursos Altamente Migratorios, 2016. Convenio de Desempeño 2016*; Informe Final, Subsecretaría de Economía y EMT; Instituto de Fomento Pesquero: Valparaíso, Chile, 2017.
24. Alheit, J.; Bernal, P. Effects of physical and biological changes on the biomass yield of the Humboldt Current ecosystem. In *Large Marine Ecosystems—Stress, Mitigation and Sustainability*; American Association for the Advancement of Science: Washington, DC, USA, 1993; pp. 53–68.
25. Oerder, V.; Bento, J.P.; Morales, C.E.; Hormazabal, S.; Pizarro, O. Coastal Upwelling Front Detection off Central Chile (36.5–37°S) and Spatio-Temporal Variability of Frontal Characteristics. *Remote Sens.* **2018**, *10*, 690. [CrossRef]
26. Lenart, A.; Missov, T.I. Goodness-of-fit tests for the Gompertz distribution. *Commun. Stat. Theor. Meth.* **2016**, *45*, 2920–2937. [CrossRef]
27. Bondon, P. Estimation of autoregressive models with epsilon-skew-normal innovations. *J. Multivar. Anal.* **2009**, *100*, 1761–1776. [CrossRef]
28. Azzalini, A. A Class of Distributions which includes the Normal Ones. *Scand. J. Stat.* **1985**, *12*, 171–178.
29. Henze, N. A probabilistic representation of the 'skew-normal' distribution. *Scand. J. Stat.* **1986**, *13*, 271–275.

© 2019 by the authors. Licensee MDPI, Basel, Switzerland. This article is an open access article distributed under the terms and conditions of the Creative Commons Attribution (CC BY) license (http://creativecommons.org/licenses/by/4.0/).

Article

On the Performance of Variable Selection and Classification via Rank-Based Classifier

Md Showaib Rahman Sarker [†], Michael Pokojovy [†] and Sangjin Kim [*,†]

Department of Mathematical Sciences, The University of Texas at El Paso, El Paso, TX 79968, USA; msarker@miners.utep.edu (M.S.R.S.); mpokojovy@utep.edu (M.P.)
* Correspondence: skim10@utep.edu
† These authors contributed equally to this work.

Received: 26 April 2019; Accepted: 14 May 2019; Published: 21 May 2019

Abstract: In high-dimensional gene expression data analysis, the accuracy and reliability of cancer classification and selection of important genes play a very crucial role. To identify these important genes and predict future outcomes (tumor vs. non-tumor), various methods have been proposed in the literature. But only few of them take into account correlation patterns and grouping effects among the genes. In this article, we propose a rank-based modification of the popular penalized logistic regression procedure based on a combination of ℓ_1 and ℓ_2 penalties capable of handling possible correlation among genes in different groups. While the ℓ_1 penalty maintains sparsity, the ℓ_2 penalty induces smoothness based on the information from the Laplacian matrix, which represents the correlation pattern among genes. We combined logistic regression with the BH-FDR (Benjamini and Hochberg false discovery rate) screening procedure and a newly developed rank-based selection method to come up with an optimal model retaining the important genes. Through simulation studies and real-world application to high-dimensional colon cancer gene expression data, we demonstrated that the proposed rank-based method outperforms such currently popular methods as lasso, adaptive lasso and elastic net when applied both to gene selection and classification.

Keywords: gene-expression data; ℓ_2 ridge; ℓ_1 lasso; adapative lasso; elastic net; BH-FDR; Laplacian matrix

MSC: 62F03; 62F07; 62P10

---

1. Introduction

Microarrays are an advanced and widely used technology in genomic research. Tens of thousands of genes can be analyzed simultaneously with this approach [1]. Identifying the genes related to cancer and building high-performance prediction models of maximal accuracy (tumor vs. non-tumor) based on gene expression levels are among central problems in genomic research [2–4]. Typically, in high-dimensional gene expression data analysis, the number of genes is significantly larger than the sample size, i.e., $m \gg n$. Hence, it is particularly challenging to identify those genes that are relevant to cancer disease and put forth prediction models. The main problem associated with high-dimensional data ($m \gg n$) is that of overfitting or overparametrization which leads to poor generalizability from training to test data.

Therefore, various researchers apply different types of regularization methods to overcome this "curse of dimensionality" in regression and other statistical and machine learning frameworks. These regularization approaches include, for example, the ℓ_1-penalty or lasso [5], which performs continuous shrinkage and feature selection simultaneously; smoothly clipped ℓ_1-penalty or SCAD [6], which is symmetric, non-concave and has singularities at the origin to produce sparse solutions; fussed lasso [7], which imposes the ℓ_1-penalty on the absolute difference of regression coefficients in order to enforce some smoothness of coefficients; or the adaptive lasso [8], etc. Unfortunately, ℓ_1-regularization sometimes perform inconsistently when used for variable selection [8]. In some situations, it introduces a major bias in estimated parameters in the logistic regression [9,10]. In contrast, the elastic net regularization procedure [11] as a combination of ℓ_1- and ℓ_2-penalties can successfully handle the highly correlated variables which are grouped together. Among the procedures mentioned above, elastic net and fussed lasso penalized methods are appropriate for gene expression data analysis. Unfortunately, when some prior knowledge needs to be utilized, e.g., when studying complex diseases such as cancer, those methods are not appropriate [4]. To account for a regulatory relationship between the genes and a priori knowledge about these genes, network-constrained regularization [4] is known to perform very well by incorporating a Laplacian matrix into the ℓ_2-penalty from the enet procedure. This Laplacian matrix represents a graph-structure of genes which are linked with each other. To select significant genes in high-dimensional gene expression data for classification, the graph-constrained regularization method is extended to logistic regression model [12].

Using penalized logistic regression methods [12,13] and graph-constrained procedures [4,12], we would build rank-based logistic regression method with variable screening procedure to improve the power of detecting most promising variables as well as classification capability.

The rest of this article is organized as follows. In Section 2, we describe variable screening procedure with adjusted p-values and regularization procedure for grouped and correlated predictors and present the computational algorithm. Further, we state the ranking criteria of four models and summarize the result of ranking procedure. In Section 3, we compare the proposed procedure with existing cutting edge regularization methods on simulation studies. Next, we apply four penalized logistic regression methods to the high dimensional gene expression data of colon cancer carcinoma to evaluation and comparison of the performance. Finally, we present a brief discussion of results and future research direction.

2. Materials and Methods

2.1. Adjusted p-Values: Benjamini and Hochberg False Discovery Rate (BH-FDR)

Multiple hypohtesis testing methods have been playing an important role in selecting most promising features while controlling type I error in high-dimensional settings. One of the most popular methods is BH-FDR [14,15] which is concerned with the expected proportion of incorrect number of rejections among a total number of rejections. The formula is mathematically expressed as $E\left(\frac{V}{R} \middle| R > 0\right)$, where V is the number of false positives and R is the total number of rejections. In this paper, the FDR method is used both for the purpose of prelimary variable screening both in the simulation studies and real data analysis to be presented later. The procedure of the method is as follows:

(1) Let $p_1, p_2, \ldots, p_m$ be the p-values of m hypothesis tests and sort them with the increasing oder: $p_{(1)}, p_{(2)}, \ldots, p_{(m)}$.
(2) Let $\hat{i} = \max\{i \mid p_{(i)} \leq \frac{iq}{m}, i = 1, \ldots, m\}$ for a given threshold q. If $\hat{i} > 1$, then reject the null hypotheses associated with $p_{(1)}, p_{(2)}, \ldots, p_{(\hat{i})}$. Otherwise, no hypotheses are rejected.

2.2. Regularized Logistic Regression

In the following, we present the regularized logistic regression model used in this paper (cf. [12]). Since this model is an integral part of our computational algorithm to be outlined in the section to follow, presenting the formula with all appropriate notations is necessary for our purposes.

Let the $n \times (m+1)$ matrix

$$X = \begin{pmatrix} 1 & x_{11} & x_{12} & \cdots & x_{1j} & \cdots & x_{1m} \\ 1 & x_{21} & x_{22} & \cdots & x_{2j} & \cdots & x_{2m} \\ \vdots & \vdots & \vdots & \ddots & \vdots & \ddots & \vdots \\ 1 & x_{i1} & x_{i2} & \cdots & x_{ij} & \cdots & x_{2m} \\ \vdots & \vdots & \vdots & \ddots & \vdots & \ddots & \vdots \\ 1 & x_{n1} & x_{n2} & \cdots & x_{nj} & \cdots & x_{nm} \end{pmatrix}$$

denote the design matrix, where n is the sample size and m is the total number of predictor variables. Without loss of generality, we assume the data are standardized with respect to each variable. This step is also performed by the pclogit R-package used in the present paper. Define the parameter vector $\eta = (\beta_0, \beta)$ comprised of an intercept β_0 and m "slopes", $\beta_1, \ldots, \beta_m$. The objective function then is written as

$$f(\eta) = -L(\eta) + p(\beta) \tag{1}$$

with the log-likelihood function

$$L(\eta) = \frac{1}{n} \sum_{i=1}^{n} [y_i \log \pi(x_i) + (1 - y_i) \log(1 - \pi(x_i))]$$

and resulting probabilities

$$\pi(x_i) = \frac{\exp(\beta_0 + x_i^T \beta)}{1 + \exp(\beta_0 + x_i^T \beta)}.$$

Here, $p(\beta)$ is the penalty function and the response variable y_i takes the value 1 for cases and 0 for controls. The i-th individual is deemed case or control based on the probability π_i. Following [4], statistical dependence among the m explanatory variables can be modeled by a graph, which, in turn, can be described by its m-dimensional Laplacian matrix $L = (L(u,v) \mid u,v \text{ vertices})$ with the entries

$$L(u,v) = \begin{cases} 1, & \text{if } u = v \text{ and } d_u \neq 0, \\ -(d_u d_v)^{-\frac{1}{2}}, & \text{if } u \text{ and } v \text{ are adjacent}, \\ 0, & \text{otherwise}. \end{cases}$$

Here, d_v is the degree of a vertex v, i.e., the number of edges through this vertex. If there is no link in v (i.e., v is isolated), then $d_v = 0$. The martix L is symmetric, positive semi-definite and has 0 as the smallest eigenvalue and 2 as the largest eigenvalue. In the following, we will write $u \sim v$ to refer to adjacent vertices. The penalty term in equation (1) can is defined as

$$p(\beta) = \lambda_1 \|\beta\|_1 + \lambda_2 \beta^T L \beta = \lambda_1 \sum_{j=1}^{m} |\beta_j| + \lambda_2 \sum_{u=1}^{m} \sum_{u \sim v} \left(\frac{\beta_u}{\sqrt{d_u}} - \frac{\beta_v}{\sqrt{d_v}} \right)^2. \tag{2}$$

Here, λ_1 and λ_2 are tuning parameters meant to control the sparsity and smoothness, $\|\beta\|_1$ is the ℓ_1-norm and $\sum_{u \sim v}(\ldots)$ denotes the summation over all adjacent vertex pairs. When $\lambda_2 = 0$, the penalty reduces to that of lasso [5], and if L is replaced by the $m \times m$-identity matrix I, the penalty corresponds to that of an elastic net [11]. If $\lambda_1 = 0$ and $L = I$, we arrive at ridge regression. In Equation (2), the penalty consists of ℓ_1- and ℓ_2-components. The ℓ_2-penalty is a degree-scaled difference of coefficients between linked predictors. According to [4], the predictor variables with more connections have larger coefficients. That is why small change of expression in the variables can lead to large change in response. Thus, this imposes sparsity and smoothness as well as correlation and grouping effects among variables. In case-control DNA methylation data analysis, ring networks and fully connected networks (cf. Figure 1) are typically used to describe correlation pattern of CpG sites within genes [12]. The Laplacian matrix is sparse and tri-diagonal (except for two corner elements) for ring networks and has all non-zero elements for fully connected networks. Those variables with more links produce strong grouping effects and are more likely to be selected in both networks [12].

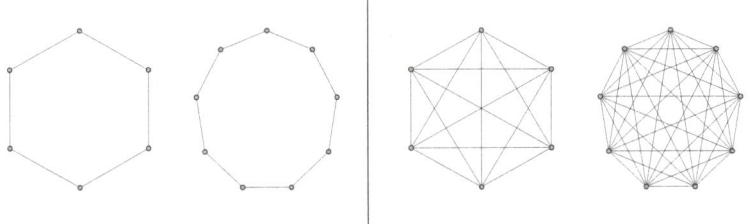

Figure 1. The ring network (left) and F.con network (right) are shown for the case there are two genes consisting of 6 and 9 CpG sites, respectively.

2.3. Computational Algorithm

Li & Li (2010) [16] developed an algorithm for graph-constrained regularization motivated by a coordinate descent algorithm from [17] for solving the unconstrained minimization problem for the objective in Equation (1). The algorithm implementation from the pclogit R-package [12,13] replaced the identity matrix by Laplacian matrix in the elastic net algorithm from the glmnet R-package [18]. According to Equation (1), the objective function is

$$f(\eta) = -L(\eta) + p(\beta),$$

where

$$p(\beta) = \lambda \alpha \sum_{i=1}^{m} |\beta_i| + \frac{1}{2}\lambda(1-\alpha) \sum_{u=1}^{m} \sum_{u \sim v} \left(\frac{\beta_u}{\sqrt{d_u}} - \frac{\beta_v}{\sqrt{d_v}} \right)^2 \qquad (3)$$

with $\lambda = \lambda_1 + 2\lambda_2$ and $\alpha = \frac{\lambda_1}{\lambda_1 + 2\lambda_2}$ for some $\lambda_1, \lambda_2 > 0$.

Following [18], we perform a second-order Taylor expansion of $L(\cdot)$ around the current estimate (β_0^*, β^*) to approximate the objective $L(\cdot)$ in Equation (1) via

$$f^*(x) = -\frac{1}{2n} \sum_{i=1}^{n} q_i (t_i - \beta_0 - x_i^T \beta)^2 + p(\beta),$$

where

$$t_i = \beta_0^* + x_i^T \beta^* + q_i^{-1}(y_i - \pi^*(x_i)),$$
$$q_i = \pi^*(x_i)(1 - \pi^*(x_i)),$$
$$\pi^*(x_i) = 1 - \left(1 + \exp(\beta_0 + x_i^T \beta^*)\right)^{-1}.$$

Now, if all other estimates for all $v = u$ are fixed, $\beta_u = \beta_u^*$ can be computed. To update the estimate from β_u^*, we have to set the gradient of $f^*(\cdot)$ equal zero (strictly speaking, zero has to be included in the subgradient of $f^*(\cdot)$) and then solve for β_u to obtain

$$\beta_u^* = \frac{S\left(\frac{1}{n} \sum_{i=1}^{n} q_i x_{iu}(t_i - t_i^{(\tilde{u})}) + \lambda(1-\alpha)g(u), \lambda\alpha\right)}{\frac{1}{n} \sum_{i=1}^{n} q_i x_{iu}^2 + \lambda(1-\alpha)},$$

where

$$t_i^{(\tilde{u})} = \beta_0^* + \sum_{j \neq u} x_{ij}\beta_j^*,$$

$$g(u) = \sum_{u \sim v} \frac{\beta_v^*}{\sqrt{d_u d_v}} \qquad (4)$$

and $s(z, r)$ denotes the "soft threshholding" operator given by

$$s(z, r) = \text{sign}(z)(|z| - r)_+ = \begin{cases} z - r, & \text{if } z > 0 \text{ and } r < |z|, \\ z + r, & \text{if } z < 0 \text{ and } r < |z|, \\ 0, & \text{otherwise}. \end{cases}$$

If the u-th predictor has no links to other predictors, then $g(u)$ in Equation (4) becomes zero, while Equation (3) takes the form

$$p(\beta) = \lambda \alpha \sum_{i=1}^{m} |\beta_i| + \frac{1}{2}\lambda(1-\alpha) \sum_{u=1}^{m} \beta_u^2.$$

Thus, the regularization reduces to that of the elastic net (enet) procedure. In general, when the linkage is nontrivial, the term $\lambda(1-\alpha)g(u)$ is added to the elastic net to get the desired grouping effect.

2.4. Adaptive Link-Constrained Regularization

When there is a link between two predictors but their regression coefficients have different signs, the coefficients cannot be expected to be smooth [16]—even locally. To resolve this problem, we first need to estimate the sign of the coefficients and then refit the model with estimated signs. When the number of predictor variables is smaller than that of sample points, ordinary least squares are performed, while ridge estimates are computed, otherwise. We have to modify the Laplacian matrix in the penalty function:

$$L^*(u, v) = \begin{cases} 1, & \text{if } u = v \text{ and } d_u \neq 0, \\ -s_u s_v (d_u d_v)^{-\frac{1}{2}}, & \text{if } u \text{ and } v \text{ are adjacent}, \\ 0, & \text{otherwise} \end{cases}$$

and then update the $g(\cdot)$-function in Equation (4) via

$$g^*(u) = \sum_{u \sim v} \frac{s_u s_v \beta_v^*}{\sqrt{d_u d_v}}.$$

2.5. Accuracy, Sensitivity, Specificity and Area under the Receiver Operating Curve (AUROC)

We evaluated four metrics of binary classification for each of lasso, adaptive lasso, elastic net and the proposed rank based logistic regression methods to compare the performance. These metrics are accuracy, sensitivity, specificity and AUROC.

Based on the notations in Table 1, we define

$$\text{Accuracy} = \frac{a+b}{a+b+c+d}, \qquad \text{Specificity} = \frac{d}{n-m}, \qquad \text{Sensitivity} = \frac{a}{m}$$

as well as

$$\text{TPR (true positive rate)} = \frac{a}{k}, \qquad \text{FPR (false positive rate)} = \frac{b}{n-m}.$$

The last metric AUROC is related to the probability that the classifier under consideration will rank a randomly selected positive case higher than a randomly selected negative case [19]. The values of all these fours metrics—accuracy, sensitivity, specificity and AUROC—range from 0 to 1. The value of 1 represents a perfect model whereas the value of 0.5 corresponds to "coin tossing". The class prediction for each individual in binary classification is made based on a continuous random variable z. Given a threshold k as a tuning parameter, an individual is classified as "positive" if $z > k$ and "negative", otherwise. The random variable z follows a probability density $f_1(z)$ if the individual belongs to "positives" and $f_0(z)$, otherwise. So, the true positive and true negative rates are given by

$$\text{TPR}(k) = \int_k^\infty f_1(z)dz \quad \text{and} \quad \text{FPR}(k) = \int_k^\infty f_0(z)dz, \text{ respectively.}$$

Now, the AUROC statistic can be expressed as

$$A = \int_0^1 \text{TPR}(\text{FPR}^{-1}(z))dz = \int_{-\infty}^\infty \int_{-\infty}^\infty \mathbb{1}\{k' > k\} f_1(k') f_0(k) dk' dk = \mathbb{P}(z_1 > z_0),$$

where z_1 and z_0 are the values of positive or negative instances, respectively.

Table 1. Confusion table: a is the number of true positives, b the number of false positives, c the number of false negatives and d the number of true negatives.

Predicted Condition	True Condition		
	Positive	Negative	Total
Positive	a	b	k
Nnegative	c	d	$n - k$
Total	m	$n - m$	$n = a + b + c + d$

2.6. Ranking and Best Model Selection

The penalty function in Equation (3) has two tuning parameters, namely, $\alpha \in [0,1]$ and $\lambda > 0$. The "limiting" cases $\alpha = 0$ and $\alpha = 1$ correspond to ridge and lasso regression, respectively. For a fixed value of α, the model selects more variables for smaller λ's and fewer variables for larger λ's. Theoretically, the result continuosly depends on α and should not significantly change under small perturbations of the latter [12,13]. Empirically, however, we discovered that the results produced by pclogit significantly vary with α. In pclogit, the Laplacian matrix determines the group effects of predictors and is calculated from adjacency matrix via

$$L = D - A,$$

where D is the degree matrix and A is the adjacency matrix. The degree-scaled difference of predictors in Equation (3) is computed from the normalized Laplacian matrix

$$L = I - D^{-\frac{1}{2}} A D^{-\frac{1}{2}}.$$

We computed the adjacency matrix by using the information from the correlation matrix obtaining

$$A(u,v) = \begin{cases} 1, & \text{if } u \neq v \text{ and } |\text{cor}(u,v)| \geq \epsilon, \\ 0, & \text{if } u = v \text{ or } |\text{cor}(u,v)| < \epsilon. \end{cases}$$

Here, $\epsilon \in (0,1)$ is a specific cut-off value for correlation. So, ϵ is another tuning parameter in our model which needs to be optimally selected. In summary, to find an optimal combination of parameters α and ϵ, we make the combination of tuning parameter α and ϵ, where the total number of combinations is given by

$$C = K \times L$$

with K and L being the number of ϵ and α values, respectively. We compared the performance for each of different combinations with T resamplings. The (negative) measure of performance for each combination is the misclassification or error rate. The pair (α, ϵ) producing the smallest misclassification rate is declared optimal and used in the next step. The sparse coefficient matrix with dimensions $m \times$ nlam (nlam = number of λ's) is used in pclogit (cf. [12,13]). By default, nlam = 100. We extracted all predictors with non-zero coefficients for each of λ values. Then we built 100 logistic regression models. Given estimated parameter values β, we have the estimated class probability for a predictor vector x at each of λ values.

$$\pi(x) = \frac{\exp(x^T \beta)}{1 + \exp(x^T \beta)}$$

Using the "naïve" Bayesian approach, we infer $y = 1$ if $\pi \geq 0.5$ and $y = 0$, otherwise. The values of accuracy, sensitivity, specificity and AUROC statistics are computed for each of 100 models and ranked in an increasing order by their values. Note that AUROC method does not use a fixed cut-off value, e.g., $c = 0.5$, but rather describes the overall performance with all possible cut-off values in the decision rule. Let R_{ij}, $i = 1, 2, 3, 4$, $j = 1, 2, \ldots, 100$, comprise the ranking matrix R. The first row, i.e., $i = 1$, displays the ranking of models with respect to their accuracy. Similarly, $i = 2$ ranks the models with respect to their sensitivity, $i = 3$, in terms of specificity and $i = 4$ by AUROC. Suppose, $R_{1,5} > R_{1,8}$. Then in the 1st row (i.e., in terms of accuracy), model 5 outperforms model 8. We calculate the column means $(\bar{R}_{\cdot j})$ of the R matrix. The column with the highest overall mean value of accuracy, sensitivity, specificity and AUROC will be chosen as the resulting optimal model. Note that there is a one-to-one correspondence between columns and the 100 competing models. In (the unlikely) case of two or more columns producing the same mean, the column with a smaller index j is selected since the model represented by such column is more parsimonious. Formally, suppose p and q, $p > q$, are two column indices in the R matrix. If $\bar{R}_{\cdot p} = \bar{R}_{\cdot q} = \max_r \bar{R}_{\cdot r}$, the q-th column will be selected and the associated model becomes our proposed rank-based penalized logistic regression model.

3. Results

3.1. Analysis of Simulated Data

We conducted extensive simulation studies to compare the performance in terms of accuracy, sensitivity, specificity and AUROC as well as the power of detecting true important variables by the proposed method with the performance of such three prominent regularized logistic regression methods as lasso, adaptive lasso and elastic net. We decided to focus on these (meanwhile) classical methods due to their popularity both in the literature and applications. Some of their very recently developed comptetitors such as [20] (R-package SelectiveInference) and [21] (R-package islasso) are currently gaining attention from the community and will be used as benchmarks in our future research.

Continuing with the description of our simulation study, all predictors x were generated from a multivariate normal distribution with the following probability density function

$$f(x) = \left(\frac{1}{2\pi}\right)^{\frac{m}{2}} \frac{1}{\sqrt{\det(\Sigma)}} \exp\left(-\frac{1}{2}(x-\mu)^T \Sigma^{-1} (x-\mu)\right)$$

with an m-dimensional mean vector μ and an $(m \times m)$-dimensional covariance matrix Σ. Writing out the covariance matrix

$$\Sigma = (\sigma_{ij}) = \begin{pmatrix} \sigma_{11} & \sigma_{12} & \sigma_{13} & \cdots & \sigma_{1m} \\ \sigma_{21} & \sigma_{22} & \sigma_{23} & \cdots & \sigma_{2m} \\ \vdots & \vdots & \vdots & \ddots & \vdots \\ \sigma_{m1} & \sigma_{m2} & \sigma_{m3} & \cdots & \sigma_{mm} \end{pmatrix} \quad \text{with} \quad \sigma_{ii} = \sigma_i^2,$$

the correlation matrix M can be expressed as

$$M = (\rho_{ij}) = \begin{pmatrix} \rho_{11} & \rho_{12} & \rho_{13} & \cdots & \rho_{1m} \\ \rho_{21} & \rho_{22} & \rho_{23} & \cdots & \rho_{2m} \\ \vdots & \vdots & \vdots & \ddots & \vdots \\ \rho_{m1} & \rho_{m2} & \rho_{m3} & \cdots & \rho_{mm} \end{pmatrix} \quad \text{with} \quad \rho_{ij} = \frac{\sigma_{ij}}{\sqrt{\sigma_{ii}^2 \sigma_{jj}^2}}.$$

The binary response variable is generated using Bernoulli distribution with individual probability (π) defined as

$$\pi(x) = \frac{1}{1 + \exp(-x^T \beta)},$$

x is the matrix of true important variables and β is the associated preassigned regression coefficients. Next, we present the details of the three different simulation scenarios considered.

- Under scenario 1, each of the simulated datasets has 200 observations and 1000 predictors. Here, for all x vectors, we let $\mu = 0$ and $\mathbf{Var}(x_j) = 0.3$. Pairwise correlation of $\rho = 0.4$ was applied to the first eight variables, while the remaining 992 variables were left uncorrelated. The β-vector was chosen as

$$\beta = (\underbrace{2,2,2,2,2}_{5 \text{ entries}}, \underbrace{3,3,3}_{3 \text{ entries}}, \underbrace{0,0,0,\ldots,0}_{992 \text{ entries}}).$$

Each of the datasets was split into training and test sets with equal proportions.

- The datasets under scenario 2 also have 200 observations and 1000 predictors. Again, $\mu = 0$ and $\mathbf{Var}(x_j) = 0.3$. Now, the first five variables were assumed to have a correlation of $\rho = 0.4$. The remaining 995 variables were independent. The β-vector was selected as

$$\beta = (\underbrace{2.0, 2.0, 2.0, 2.7, 2.0, 2.0, 2.5, 2.7, -2.8, 3.0, 2.6, 3.0, 3.0, 3.0, 3.0}_{15 \text{ entries}}, \underbrace{0,0,0,\ldots,0}_{985 \text{ entries}}).$$

Each of the datasets was split into training and test sets with equal proportions.

- Under the last scenario 3, each of the datasets has 150 observations and 1000 predictors. We let $\mu = 0$ and $\mathbf{Var}(x_j) = 0.4$. The first five variables were assigned into a correlation value of $\rho = 0.3$, while the variables with indices from 11 to 30 were chosen to have the correlation value of $\rho = 0.6$. Outside of these two blocks, the variables were assumed uncorrelated. The β-vector was chosen

$$\beta = (\underbrace{2.0, 2.0, 2.0, 2.0, 2.0, 2.5, -2.6, 2.7, 3.0, -2.9, 2.0, 2.0, 2.0, 2.0, 2.0,}_{15 \text{ entries}}$$
$$\underbrace{2.5, -2.0, 2.7, 3.0, -2.5, 2.0, 2.0, 2.0, 2.0, 2.0, 2.5, -2.0, 2.7, 3.0, -2.5,}_{15 \text{ entries}} \underbrace{0,0,0,\ldots,0}_{970 \text{ entries}}).$$

The dataset was split into training and test sets with ratio of 70 to 30.

We compared the proposed rank-based penalized logistic regression method with lasso, adaptive lasso and elastic net methods from the glmnet R-package [11]. Algorithm 1 summarizes the procedure to calculate the average value of accuracy, sensitivity, specificity and AUROC based on a given number of iterations for each of the three simulation scenarios.

Algorithm 1 Calculation of overall mean and standard deviation on simulation studies
---
Step 1: Generate the data on each of the three simulation scenarios.
Step 2: Split the data into training and test sets randomly with the ratio of 70 to 30.
Step 3: Screen the variables using BH-FDR based on the training dataset.
Step 4: Plug the screened variables to each of the four methods.
Step 5: Calculate the values of Accuracy, Sensitivity, Specificity and AUROC for each of the methods.
Step 6: Repeats Step 1–5 to achieve a given number of replications.
Step 7: Calculate the means and standard deviations for each of the methods.

In Table 2, we compare the estimated mean and standard deviation of accuracy, sensitivity, specificity and AUROC values based on 200 iterations under correlation structure of $\rho = 0.4$ in the simulation of scenario 1. The proposed rank-based penalized method shows the highest accuracy of 0.963 with the standard deviation of 0.02, sensitivity of 0.961 with standard deviation of 0.03, specificity of 0.965 with standard deviation of 0.03. In addition, it yields the same AUROC of 0.995 with standard deviation of 0.01 as elastic net and adaptive lasso.

Table 2. Comparison of the performance among the four methods over 200 replications under simulation scenario 1. The values in parentheseses are the standard deviations.

Method	Accuracy	Sensitivity	Specificity	AUROC
rank-based	0.963 (0.02)	0.961 (0.03)	0.965 (0.03)	0.995 (0.01)
lasso	0.953 (0.03)	0.952 (0.04)	0.955 (0.03)	0.993 (0.01)
alasso	0.957 (0.03)	0.955 (0.04)	0.960 (0.03)	0.995 (0.01)
enet	0.961 (0.02)	0.959 (0.04)	0.962 (0.03)	0.995 (0.01)

In Table 3, we compare estimated mean and standard deviation of accuracy, sensitivity, specificity and AUROC values using 200 iterations under correlation structure of $\rho = 0.4$ in the simulation of scenario 2. The proposed rank-based method also shows highest accuracy of 0.831 with standard deviation 0.04, sensitivity of 0.833 with standard deviation of 0.06, specificity of 0.829 with standard deviation of 0.05. In addition, the proposed method produces AUROC of 0.913 with standard deviation of 0.03. This is the second highest value which is slightly lower than the AUROC value of the elastic net.

Table 3. Comparison of the performance among the four methods over 200 replications under simulation scenario 2. The values in parentheseses are the standard deviations.

Method	Accuracy	Sensitivity	Specificity	AUROC
rank-based	0.831 (0.04)	0.833 (0.03)	0.829 (0.05)	0.913 (0.03)
lasso	0.826 (0.05)	0.827 (0.07)	0.825 (0.09)	0.910 (0.04)
alasso	0.815 (0.04)	0.814 (0.07)	0.815 (0.07)	0.902 (0.04)
enet	0.826 (0.04)	0.828 (0.07)	0.825 (0.07)	0.915 (0.03)

In Table 4, we compare estimated mean and standard deviation of accuracy, sensitivity, specificity and AUROC values with 150 iterations under correlation structure of $\rho = 0.3$ and $\rho = 0.6$ in simulation of scenario 3. The proposed method shows highest accuracy of 0.916 with standard deviation of 0.04, sensitivity of 0.919 with standard deviation of 0.06, specificity of 0.912 with standard deviation of 0.06 and AUROC of 0.977 with standard deviation of 0.02.

Table 4. Comparison of the performance among the four methods over 150 replications under simulation scenario 3. The values in parentheseses are the standard deviations.

Method	Accuracy	Sensitivity	Specificity	AUROC
rank-based	0.916 (0.04)	0.919 (0.06)	0.912 (0.06)	0.977 (0.02)
lasso	0.888 (0.04)	0.898 (0.06)	0.880 (0.07)	0.963 (0.02)
alasso	0.866 (0.04)	0.877 (0.07)	0.855 (0.07)	0.949 (0.03)
enet	0.909 (0.04)	0.916 (0.06)	0.903 (0.06)	0.975 (0.02)

Furthermore, we compared the performance in terms of selecting the number of true important variables by each of the four methods under three different simulation scenarios. First, we performed multiple hypothesis testing with BH-FDR [15] to reduce the dimensionality of the data. After performing a screening step to retain the relevant variables, we used them as input for the proposed rank-based penalized method with the regularization step outlined in Section 2.3. We illustrate the performance of variable selection with boxplots in Figures 2–4 for simulation scenarios 1, 2, and 3. Each figure displays two boxplots, which, in turn, depict the distribution of the number of variables selected (NVS) and the number of true important variables (NTIV) within the number of variables selected (NVS) with each of the four methods computed based on the given number of iterations in each of the three simulation scenarios.

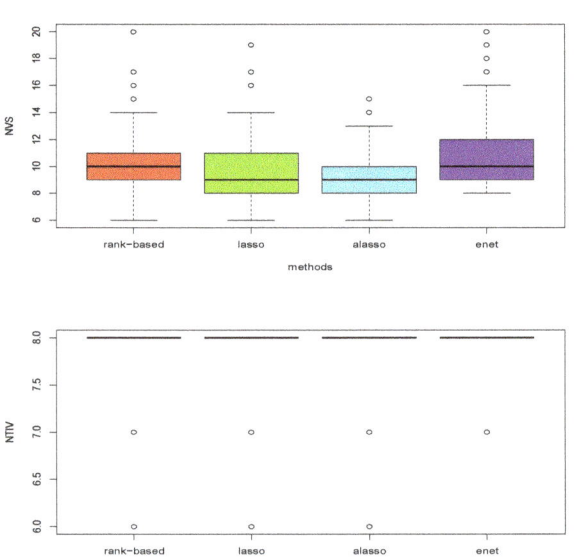

Figure 2. Boxplots of total number of variables (NVS) selected and the number of true important variables (NTIV) within the number of variables selected with four different models under scenario 1 based on 200 replications.

Figure 2 reports that the proposed rank-based method has a slightly higher median number of variables selected (displayed as a thick line in the upper boxplots) than lasso, adaptive lasso and elastic net under scenario 1. The lower boxplots show that all four methods performed head-to-head for selection of true important variables under scenario 1 with 200 replications. Table 5 compares the mean and the standard deviation (in parentheseses) of the number of variables (NVS) selected and the number of true important variables (NTIV) in NVS for each of the four methods over 200 replications. The proposed rank-based method and elastic net performed head-to-head while slightly outperforming lasso and adaptive lasso.

Table 5. Estimated mean and standard deviation of number of variables (NVS) selected and the number of true important variables (NTIV) among NVS with four different models under simulated scenario 1 with 200 replications. The values in parentheseses are standard deviations.

Method	NVS	NTIV
rank-based	10.465 (2.24)	7.975 (0.19)
lasso	9.885 (1.98)	7.880 (0.37)
alasso	9.475 (1.47)	7.970 (0.20)
enet	10.805 (2.37)	7.975 (0.16)

Figure 3 suggests the proposed method has a marginally higher median number of variable selected compared to the other three methods in the upper boxplot. It is also clear that the proposed method has a slightly higher median number of true important variables in the lower boxplot on scenarios 2 computed with 200 replications. Table 6 confirms that the rank-based penalized method has the highest mean both for selecting the number of variables and important variables.

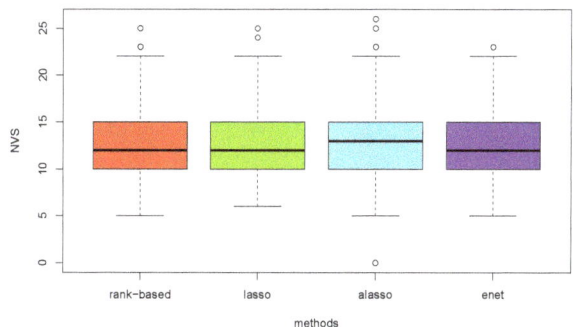

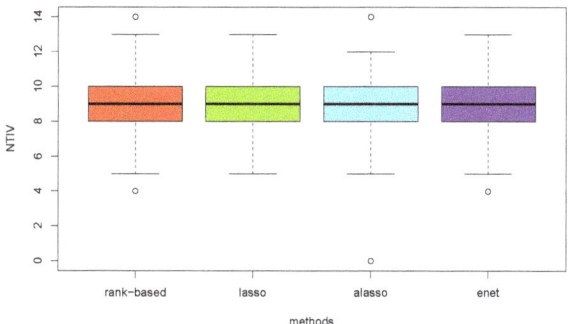

Figure 3. Boxplots of total number of variables (NVS) selected and the number of true important variables (NTIV) within the number of variables selected with four different models on scenario 2 based on 200 replications.

Table 6. Estimated mean and standard deviation of number of variables (NVS) selected and the number of true important variables (NTIV) among NVS in four different models under simulation scenario 2 with 200 replications. The values in parentheseses are standard deviations.

Method	NVS	NTIV
rank-based	13.675 (3.95)	9.345 (1.62)
lasso	12.750 (3.50)	8.905 (1.81)
alasso	11.965 (3.18)	8.720 (1.73)
enet	13.115 (3.92)	9.105 (1.73)

In Figure 4, the upper boxplot demonstrates that the proposed rank-based method has the highest median number of variables selected, elastic net has second highest median, lasso has third largest median and adaptive lasso has the smallest median under scenario 3 based on 150 replications. The lower boxplots also show that the proposed rank-based method has the highest median number of true important variables selected. However, adaptive lasso has a higher median number of true important variables than lasso unlike the upper boxplots. Thus, the proposed rank based-method clearly outperforms other three methods under high-correlation settings among variables.

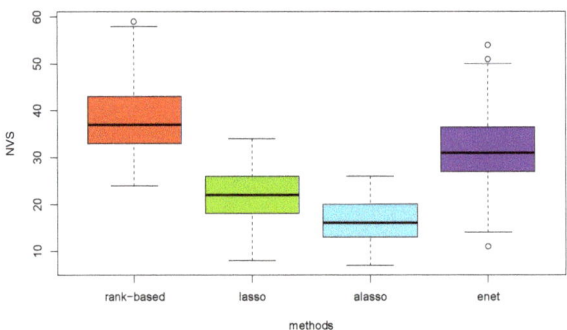

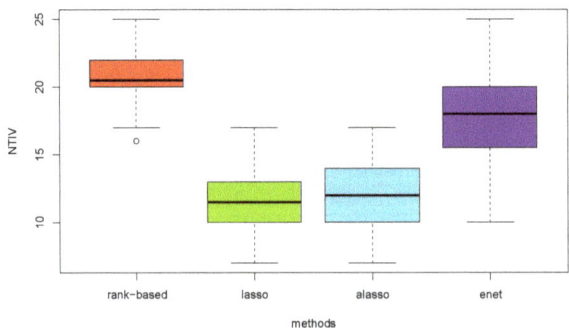

Figure 4. Boxplots of total number of variables (NVS) selected and the number of true important variables (NTIV) within the number of variables selected with four different models under scenario 3 based on 150 replications.

Table 7 summarizes the number of variables selected and true important variables selected across the four methods under the high-correlation setting among variables computed from 150 replications.

The proposed rank-based method has the highest mean number of overall variables selected and true important variables selected.

Table 7. Estimated mean and standard deviation of number of variables (NVS) selected and the number of true important variables (NTIV) among NVS in four different models under simulation scenario 3 with 150 replications. The values in parentheseses are standard deviations.

Method	NVS	NTIV
rank-based	37.830 (7.14)	20.770 (1.88)
lasso	22.010 (4.96)	11.780 (2.21)
alasso	16.430 (4.39)	11.920 (2.28)
enet	32.270 (8.55)	17.600 (3.15)

3.2. Real Data Example

We applied four logistic regression methods to select differentially expressed genes and assess their discrimination capability between colon cancer cases and healthy controls using high-dimensional gene expression data [22]. The colon cancer gene expression dataset is available at [23]. It contains 2000 genes with the highest minimal intensity across 62 tissues. The data were measured on 40 colon tumor samples and 22 normal colon tissue samples. We split the data set into training and testing sets with proportions 70% and 30%, respectively. To detect significantly differentially expressed genes for high-dimensional colon cancer carcinoma and measure classification prediction, we adapted two step procedures of filtering and variable selection. First, we applied BH-FDR [15] to select most promising candidates of genes as a preprocessing step and then used the screened genes as input to the proposed rank-based method and three other popular methods—lasso, adaptive lasso and elastic net. The performance in terms of accuracy, sensitivity, specificity and AUROC as well as the selection probabilities for the four methods are reported in Tables 8 and 9, respectively.

Algorithm 2 outlines above protocols the procedure of calculating the average values of accuracy, sensitivity, specificity and AUROC through 100 bootstrap iterations applied to the colon cancer gene expression data. In Table 8, the performance of all four metrics are computed based on 100 iterations of resampled subsets of individuals.

Algorithm 2 Calculation of mean with standard deviation on colon cancer data
---
Step 1: Split the data into training and test sets randomly with the ratio of 70 to 30.
Step 2: Screen genes with the BH-FDR method based on the training data.
Step 3: Plug the screened genes as the input to each of four methods.
Step 4: Calculate the values of Accuracy, Sensitivity, Specificity and AUROC across each of the methods on the test data.
Step 5: Repeat Steps 1 through 4 for 100 times.
Step 6: Calculate means and standard deviations for each of the methods.

Table 8. Estimated mean values and standard deviations for the four metrics across the four competing penalized logistic regression models computed from 100 resamplings. The values in parentheseses are standard deviations.

Colon Cancer Data Analysis Based on 100 Times Resmpling				
Method	Accuracy	Sensitivity	Specificity	AUROC
rank-based	0.853 (0.08)	0.860 (0.13)	0.840 (0.13)	0.917 (0.06)
lasso	0.801 (0.09)	0.911 (0.07)	0.637 (0.21)	0.897 (0.08)
adaptive lasso	0.804 (0.09)	0.869 (0.09)	0.719 (0.21)	0.877 (0.08)
elastic net	0.802 (0.09)	0.917 (0.07)	0.640 (0.22)	0.903 (0.07)

The average AUROC of 0.853 with standard deviation of 0.06 in the proposed rank-based method has the highest value compared to other three methods. Also the accuracy of 0.853 with standard deviation of 0.08 are optimal among the four methods. The values of sensitivity (0.860) and specificity (0.840) are also better than those of the other three methods. In summary, it is fair to conclude that the proposed rank-based method outperforms the other three popular penalized logistic regression methods. Table 9 shows top 5 ranked genes with highest selection probabilities for the proposed rank-based method, lasso, adaptive lasso and elastic net. An expressed sequence tag (EST) of Hsa.1660 associated with colon cancer carcinoma is found by all four methods. Hsa.36689 [24,25] is shown and top ranked by the proposed method, lasso and elastic net. Hsa692 also appeared and is second ranked by the proposed method, lasso and elastic net. In addition, Hsa.37937 is shown and is third and second ranked by the proposed method and elastic net, respectively.

Table 9. List of top 5 ranked genes across rank-based, lasso, adaptive and elastic net. An extra asterix (*) sign is put next to a gene each time the gene is selected by one of four methods.

EST Name	Gene ID	Gene Description	Selection Probability
		Rank-Based	
***Hsa.36689	Z50753	H.sapiens mRNA for GCAP-II/uroguanylin precursor	1.00
***Hsa.692.2	M76378	Human cysteine-rich protein (CRP) gene, exons 5 and 6	0.99
**Hsa.37937	R87126	Myosin heavy chain,nonmuscle(Gallus gallus)	0.97
****Hsa.1660	H55916	Peptidyl-prolyl cis-trans isomerase, mitrochondrial precursor(human)	0.91
Hsa.1832	R44887	nedd5 protein (Mus musculus)	0.90
		Lasso	
***Hsa.36689	Z50753	H.sapiens mRNA for GCAP-II/uroguanylin precursor	0.87
Hsa.692.2	M76378	Human cysteine-rich protein (CRP) gene, exons 5 and 6	0.82
*****Hsa.1660	H55916	Peptidyl-prolyl cis-trans isomerase, mitrochondrial precursor(human)	0.66
Hsa.6814	H08393	Collagen alpha 2(XI) chain(Homo sapiens)	0.52
Hsa.8147	M63391	Human desmin gene, complete cds	0.50
		Adaptive Lasso	
Hsa.1454	M82919	H. gamma amino butyric acid(GABAA)receptor beta3 subunit mRNA,cds	0.83
Hsa.6814	H08393	Collagen alpha 2(XI) chain(Homo sapiens)	0.77
****Hsa.1660	H55916	Peptidyl-prolyl cis-trans isomerase, mitrochondrial precursor(human)	0.77
Hsa.14069	T67077	Sodium/Potasssium-transporting atpase gamma chain(Ovis aries)	0.69
Hsa.2456	U25138	Human MaxiK potassium channel beta subunit mRNA, complete cds	0.55
		Elastic Net	
***Hsa.36689	Z50753	H.sapiens mRNA for GCAP-II/uroguanylin precursor	0.98
**Hsa.37937	R87126	Myosin heavy chain,nonmuscle(Gallus gallus)	0.94
***Hsa.692.2	M76378	Human cysteine-rich protein (CRP) gene, exons 5 and 6	0.94
Hsa.8147	M63391	Human desmin gene, complete cds	0.91
****Hsa.1660	H55916	Peptidyl-prolyl cis-trans isomerase, mitrochondrial precursor(human)	0.84

4. Discussion

In this paper, we proposed a new rank-based penalized logistic regression method to improve classification performance and the power of variable selection in high-dimensional data with strong correlation structure.

Our simulation studies demonstrated that the proposed method improves not only the performance of classification or class prediction but also the detection of true important variables under various correlation settings among features when compared to existing popular regularization methods such as lasso, adaptive lasso, and elastic net. As demonstrated by simulation studies, if the true important variables are not passed through the filtering method such as BH-FDR, their chance of being selected in the final model decreases signficantly, thus, leading to reduction in variable selection and classification performance. Therefore, effective filtering methods which are likely to retain as many most promising variables as possible are indispensable.

Applied to high-dimensional colon gene expression data, the proposed rank-based logistic regresson method with BH-FDR screening produced the highest average AUROC value of 0.917 with standard deviation of 0.06 and accuracy of 0.853 with standard deviation of 0.08 using 100 resampling steps. The proposed method produced a good balance between sensitivity and specificity in contrast to other methods. Elastic net demonstrated the second best peformance with an average AUROC value of 0.903 with standard deviation of 0.07. A probable reason is that elastic net accounts for group correlation effects. In addition, we compared top 5 ranked ESTs across the proposed method, lasso, adaptive lasso and elastic net [12]. They had a common EST of Hsa.1660 associated to colon cancer data. We also found that Hsa.36689 was both deemed important and top ranked by the proposed method, lasso and elastic net. This also applied to Hsa.692, which was deemed important and second top ranked by the proposed method and lasso, whereas it was only third-ranked by the elastic net. Hsa.37937 was detected by both the proposed method and the elastic net. Hence, the four ESTs mentioned appear to be promising candidate biomarkers associated with colon cancer carcinoma. The function of the genes corresponding to ESTs is summarized in Table 9.

5. Conclusions

In this study the proposed rank-based classifier demonstrated the superiority in not only classification prediction but also the power of detecting true important variables when compared to lasso, adaptive lasso, and elastic net through the extensive simulation studies. Besides, in the application of high-dimensional colon cancer gene expression data, the proposed classifier showed the best performance in terms of accuracy and AUROC among the four classifiers considered in the paper. As a future research, we would develop the methodology of variable selection and compare the performance with those of most recent competitors such as [20,21,26,27], etc.

Author Contributions: All authors have equally contributed to this work. All authors wrote, read, and approved the final manuscript.

Funding: This research received no external funding.

Acknowledgments: We would like to thank the reviewers for their valuable comments.

Conflicts of Interest: The authors declare no conflict of interest.

References

1. Houwelingen, H.C.V.; Bruinsma, T.; Hart, A.A.M.; Veer, L.J.V.; Wessels, L.F.A. Cross-validated Cox regression on microarray gene expression data. *Stat. Med.* **2006**, *25*, 3201–3216. [CrossRef] [PubMed]
2. Lofti, E.; Keshavarz, A. Gene expression microarray classification using PCA–BEL. *Comput. Biol. Med.* **2014**, *54*, 180–187.
3. Algamal, Z.Y.; Lee, M.H. Penalized logistic regression with the adaptive LASSO for gene selection in high-dimensional cancer classification. *Expert Syst. Appl.* **2015**, *42*, 9326–9332. [CrossRef]
4. Li, C.; Li, H. Network-constrained regularization and variable selection for analysis of genomic data. *Bioinformatics* **2008**, *24*, 1175–1182. [CrossRef] [PubMed]
5. Tibshirani, R. Regression shrinkage and selection via the Lasso. *J. R. Stat. Soc. B* **1996**, *58*, 267–288. [CrossRef]
6. Fan, J.; Li, R. Variable selection via nonconcave penalized likelihood and its oracle properties. *J. Am. Stat. Assoc.* **2001**, *96*, 1175–1182. [CrossRef]
7. Tibshirani, R.; Saunders, M. Sparsity and smoothness via the fused lasso. *J. R. Stat. Soc. B* **2005**, *67*, 91–108. [CrossRef]
8. Zou, H. The adaptive Lasso and its oracle properties. *J. Am. Stat. Assoc.* **2006**, *101*, 1418–1429. [CrossRef]
9. Meinshausen, N.; Yu, B. Lasso-type recovery of sparse representations for high-dimensional data. *Ann. Stat.* **2009**, *37*, 246–270. [CrossRef]
10. Huang, H.; Liu, X.Y.; Liang, Y. Feature selection and cancer classification via sparse logistic regression with the hybrid $L_{1/2+2}$ regularization. *PLoS ONE* **2009**, *11*, e0149675. [CrossRef]
11. Zou, H.; Hastie, T. Regularization and variable selection via the elastic net. *J. R. Stat. Soc. B* **2005**, *67*, 301–320. [CrossRef]

12. Sun, H.; Wang, S. Penalized logistic regression for high-dimensional DNA methylation data with case-control studies. *Bioinformatics* **2012**, *28*, 1368–1375. [CrossRef] [PubMed]
13. Sun, H.; Wang, S. Network-based regularization for matched case-control analysis of high-dimensional DNA methylation data. *Stat. Med.* **2012**, *32*, 2127–2139. [CrossRef] [PubMed]
14. Reiner, H.; Yekutieli, D.; Benjamin, Y. Identifying differentially expressed genes using false discovery rate controlling procedures. *Bioinformatics* **2003**, *19*, 368–375. [CrossRef]
15. Benjamini, Y.; Hochberg, Y. Controlling the false discovery rate: A practical and powerful approach to multiple testing. *J. R. Stat. Soc. B* **1995**, *57*, 289–300. [CrossRef]
16. Li, C.; Li, H. Variable selection and regression analysis for graph-structured covariates with an application to genomics. *Ann. Appl. Stat.* **2010**, *4*, 1498–1516. [CrossRef]
17. Friedman, J.; Hastie, T.; Hofling, H.; Tibshirani, R. Pathwise coordinate optimization. *Ann. Appl. Stat.* **2007**, *1*, 302–332. [CrossRef]
18. Friedman, J.; Hastie, T.; Tibshirani, R. Regularization paths for generalized linear models via coordinate descent. *J. Stat. Softw.* **2010**, *33*, 1–22. [CrossRef]
19. Fawcett, T. An introduction to ROC analysis. *Pattern Recognit. Lett.* **2006**, *27*, 861–874. [CrossRef]
20. Lee, J.D.; Sun, D.L.; Sun, Y.; Taylor, J.E. Exact post-selection inference, with application to the lasso. *Ann. Stat.* **2016**, *44*, 907–927. [CrossRef]
21. Cilluffo, G.; Sottile, G.; La Grutta, S.; Muggeo, V.M.R. The Induced Smoothed lasso: A practical framework for hypothesis testing in high dimensional regression. *Stat. Methods Med. Res.* **2019**, doi:10.1177/0962280219842890. [CrossRef] [PubMed]
22. Alon, U.; Barakai, N.; Notterman, D.A.; Gish, K.; Ybarra, S.; Mack, D.; Levine, A.J. Broad patterns of gene expression revealed by clustering analysis of tumor and normal colon tissues probed by oligonucleotide arrays. *Proc. Natl. Acad. Sci. USA* **1999**, *96*, 6745–6750. [CrossRef]
23. Available online: http://genomics-pubs.princeton.edu/oncology/affydata/index.html (accessed on 25 April 2019).
24. Ding, Y.; Wilkins, D. A simple and efficient algorithm for gene selection using sparse logistic regression. *Bioinformatics* **2003**, *19*, 2246–2253.
25. Li, Y.; Campbell, C.; Tipping, M. Bayesian automatic relevance determination algorithms for classfifying gene expression data. *Bioinformatics* **2002**, *18*, 1332–1339. [CrossRef]
26. Frost, H.R.; Amos, C.I. Gene set selection via LASSO penalized regression (SLPR). *Nucleic Acids Res.* **2017**, doi:10.1093/nar/gkx291. [CrossRef]
27. Boulesteix, A.L.; De, B.R.; Jiang, X.; Fuchs, M. IPF-LASSO: Integrative L_1-Penalized Regression with Penalty Factors for Prediction Based on Multi-Omics Data. *Comput. Math. Methods Med.* **2017**. [CrossRef] [PubMed]

© 2019 by the authors. Licensee MDPI, Basel, Switzerland. This article is an open access article distributed under the terms and conditions of the Creative Commons Attribution (CC BY) license (http://creativecommons.org/licenses/by/4.0/).

Article

Two-Stage Classification with SIS Using a New Filter Ranking Method in High Throughput Data

Sangjin Kim [1],* and Jong-Min Kim [2]

1. Department of Mathematical Sciences, University of Texas at El Paso, El Paso, TX 79968, USA
2. Division of Sciences and Mathematics, University of Minnesota at Morris, Morris, MN 56267, USA; jongmink@morris.umn.edu
* Correspondence: skim10@utep.edu

Received: 22 April 2019; Accepted: 27 May 2019; Published: 29 May 2019

Abstract: Over the last decade, high dimensional data have been popularly paid attention to in bioinformatics. These data increase the likelihood of detecting the most promising novel information. However, there are limitations of high-performance computing and overfitting issues. To overcome the issues, alternative strategies need to be explored for the detection of true important features. A two-stage approach, filtering and variable selection steps, has been receiving attention. Filtering methods are divided into two categories of individual ranking and feature subset selection methods. Both have issues with the lack of consideration for joint correlation among features and computing time of an NP-hard problem. Therefore, we proposed a new filter ranking method (PF) using the elastic net penalty with sure independence screening (SIS) based on resampling technique to overcome these issues. We demonstrated that SIS-LASSO, SIS-MCP, and SIS-SCAD with the proposed filtering method achieved superior performance of not only accuracy, AUROC, and geometric mean but also true positive detection compared to those with the marginal maximum likelihood ranking method (MMLR) through extensive simulation studies. In addition, we applied it in a real application of colon and lung cancer gene expression data to investigate the classification performance and power of detecting true genes associated with colon and lung cancer.

Keywords: LASSO; SCAD; MCP; SIS; elastic net; accuracy; AUROC; geometric mean

---

1. Introduction

In the last decade, high dimensional data has appeared with the development of high throughput techniques, especially in the research area of machine learning [1,2] and data mining [3,4] in biology. The possibility of finding novel true important variables has potentially become high with a huge amount of data. However, due to limitations of computing capabilities and overfitting issues, two-stage approaches of filtering and variable selection for prediction purpose has been popular. These include methods for microarray [5–8] and RNA-Seq [9,10] data, and genome-wide association studies (GWAS) [11,12]. Filtering methods, which reduce dimensionality and try to retain the most promising features as possible, have long been under development. A number of filtering methods has been proposed to rank features, such as Information gain [13], Markov blanket [14], Bayesian variable selection [15], Boruta [16], Fisher score [17], Relief [18], maximum relevance and minimum redundancy (MRMR) [19], marginal maximum likelihood score (MMLs) [20], among which MMLS is one of the simplest and computationally efficient methods of feature selection with some criteria.

Feature selection methods are divided into two categories of marginal feature ranking and feature subset selection considering relationship among features. Marginal feature ranking methods order individual features by their scores and then drop out irrelevant features with small scores using the desired criteria. [21] utilized the Relief statistical method to rank features. [20] gave a marginal maximum likelihood estimator as a feature ranking method and improved classification accuracy. [22]

also developed a novel method to rank features and then chose the optimal subset of features. Individual ranking methods have been widely used in high throughput data analysis because of their simplicity and computational time efficiency but a predetermined threshold is required before variable selection stage. To overcome this issue, the sure independent screening (SIS) approach [23] was developed to ensure that all true important variables survive after the variable screening with probability tending to one. Feature subset selection methods [24–26] detect an optimal subset of features leading to the best performance of prediction. However, these methods have heavy computational time leading to be NP-hard [27] under a high dimensional setting.

In this paper, we proposed a filter ranking method (PF) utilizing selection probability with an elastic net based on resampling technique with SIS. The selected features are then applied to three popular variable selection algorithms such as least absolute shrinkage and selection operator (LASSO) [28], minimax concave penalty (MCP) [29], and smoothly clipped absolute deviation (SCAD) [30].

The rest of this article is organized as follows. In Section 2, we described three penalized logistic methods of LASSO, MCP, and SCAD, marginal maximum likelihood ranking method, sure independence screening method (SIS), the proposed statistical methods for filter ranking and its algorithm, and metrics of performance including accuracy, area under the receiver operating characteristic (AUROC), and geometric mean of sensitivity and specificity (G-mean). In Section 3, we describe the superior performance of our proposed method compared to an individual ranking method of marginal maximum likelihood logistic regression (MMLR) with SIS through the extensive simulation studies. We next applied the proposed method to the high dimensional colon gene expression data and investigate the biological meaning of selected genes. Finally, in Section 4, we discuss our findings.

2. Materials and Methods

We split this section into several subsections describing the methods used in the study. The section of sparse logistic regression, such as LASSO, adaptive LASSO, SCAD, and MCP, is discussed. A filtering method with SIS used as a reference is briefly described and then our proposed method is explained in detail. The final section considers metrics of the performance including accuracy, AUROC, and G-mean. All simulations and real applications were done with R software and the corresponding codes, results, and data are available at [31].

2.1. Penalized Logistic Regression Method

Binary logistic regression is widely used in the classification of clinical outcomes of cancer using gene expression data to identify the relationship between the outcome and a set of predictors to build prediction models. However, the logistic regression has limited use in high dimensional settings when N << P because the inverted matrix does not exist for the estimation of regression coefficients. Embedded methods such as LASSO, SCAD, and MCP are the most popular methods in gene selection under a high dimensional setting because they are allowed to select a sparse subset of genes by continuously shrinking unimportant covariates' regression coefficients into zero. A number of penalty based embedded methods has been extensively studied and modified in the area of cancer genes selection under high throughput data [32–43].

Let the expression levels of genes in i^{th} individual be denoted as $x_i = (x_{i1}, x_{i2}, \ldots, x_{id})$ for $i = 1, \ldots, n$ and d is a total number of genes. Given a training data set $\{(x_i, y_i)\}_{i=1}^n$ where $y_i \in (0,1)$, $y_i = 0$ indicates that i^{th} individual is in normal group and $y_i = 1$ in cancer group. Assuming that $p(x_i) = p(y_i = 1|x_i)$, the logistic regression is defined as follows:

$$\log\left(\frac{p(x_i)}{1-p(x_i)}\right) = \beta_0 + x_i\beta, \text{ where } i = 1,\ldots,n \text{ and } \beta = (\beta_1, \ldots, \beta_d)^T. \qquad (1)$$

The following formula is for the maximum log-likelihood estimator of logistic regression (MLR). $\hat{\beta}_{MLR}$ is defined as follows:

$$\hat{\beta}_{MLR} = \underset{\beta}{\operatorname{argmax}}[\log(\mathcal{L}(\beta))] = \underset{\beta}{\operatorname{argmax}}\left[\sum_{i=1}^{n}(y_i \log(p(x_i)) + (1 - y_i)\log(1 - p(x_i)))\right] \qquad (2)$$

The estimation of parameters can be calculated by maximizing the above log-likelihood function $\log(\mathcal{L}(\beta))$. The criterion for classification is that if $p(y_i = 1|x_i) \geq 0.5$, then the individual belongs to the cancer group, otherwise, normal group. The penalized logistic regression (PLR) is a combination of logistic regression with penalty function and parameters can be estimated by minimizing the log-likelihood function with penalty function as follows:

$$\hat{\beta}_{PLR} = \underset{\beta}{\operatorname{argmin}}\left[-\sum_{i=1}^{n}(y_i \log(p(x_i)) + (1 - y_i)\log(1 - p(x_i))) + p(\beta)\right], \qquad (3)$$

where $p(\beta)$ is a penalty function.

One of the most popular penalty functions is LASSO [12–18]. It forces most of the unimportant genes' regression coefficients into zero. Although it is widely used in high throughput biomedical data, it has the tendency to randomly choose one of the genes with high correlation and then throw out the rest of the genes. The estimation of regression coefficients can be done by minimizing the following likelihood:

$$\hat{\beta}_{lasso} = \underset{\beta}{\operatorname{argmin}}\left[-\sum_{i=1}^{n}(y_i \log(p(x_i)) + (1 - y_i)\log(1 - p(x_i))) + \lambda \sum_{i=1}^{d}|\beta_i|\right]. \qquad (4)$$

Another popular sparse logistic regression is SCAD with a concave penalty that complements the limitation of lasso mentioned above. To estimate parameters of regression coefficients, the following log-likelihood can be minimized:

$$\hat{\beta}_{SCAD} = \underset{\beta}{\operatorname{argmin}}\left[-\sum_{i=1}^{n}(y_i \log(p(x_i)) + (1 - y_i)\log(1 - p(x_i))) + \lambda \sum_{i=1}^{d}p_\lambda(\beta_i)\right]. \qquad (5)$$

The $p_\lambda(\beta_i)$ is

$$|\beta_j|I_{(|\beta_j|\leq \lambda)} + \left(\frac{\left\{(a^2-1)\lambda^2 - (a\lambda - |\beta_i|)_+^2\right\}I(\lambda \leq |\beta_i|)}{2(a-1)}\right), \text{ for } \lambda \geq 0 \text{ and } a > 2. \qquad (6)$$

The minimax concave penalty (MCP) is also popular as much as SCAD. The estimation of regression coefficients can be achieved by minimizing the following log-likelihood function:

$$\hat{\beta}_{MCP} = \underset{\beta}{\operatorname{argmin}}\left[-\sum_{i=1}^{n}(y_i \log(p(x_i)) + (1 - y_i)\log(1 - p(x_i))) + \lambda \sum_{i=1}^{d}p_\lambda(\beta_i)\right]. \qquad (7)$$

$p_\lambda(\beta_i)$ is written as follows.

$$\left(\frac{2a\lambda|\beta_i| - \beta_i^2}{2a}\right)I(|\beta_i| \leq a\lambda) + \left(\frac{a\lambda^2}{2}\right)I(|\beta_i| > a\lambda), \text{ for } \lambda \geq 0 \text{ and } a > 1. \qquad (8)$$

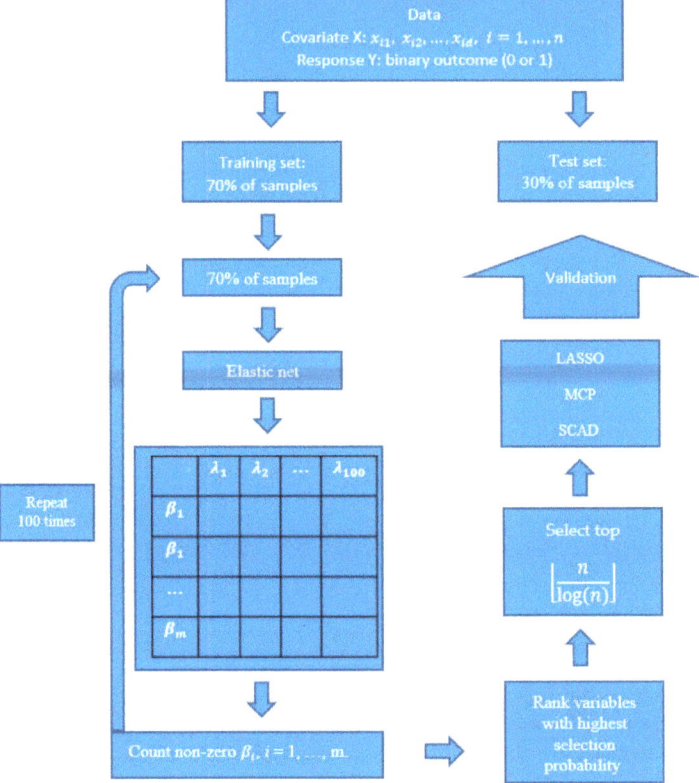

Figure 1. Diagram showing the proposed two-step procedure.

2.2. Variable Ranking with MMLR

In practice, gene expression data usually contain irrelevant genes that lead to low classification performance under high dimensional settings. Therefore, the analysis with respect to important variable detection has become a main part of the classification. Filter methods have been paid attention to such a goal. These methods essentially measure the strength of the relationship of each of genes with a binary outcome and then ranks them [44]. They have serval benefits for the analysis of huge amounts of gene expression data. First of all, they reduce high dimension into the appropriate dimension as well as the cost of computation time. Furthermore, they can also help improve the classification performance by increasing the likelihood to choose true important genes. There are a lot of filter methods applied to big data analysis of gene expression. One of the popular ranking methods is a logistic regression as a classifier. The value of maximum marginal likelihood estimator of logistic regression (MMLR) in each gene can be calculated using Equation (2) with a single gene. According to this method, a significant gene should have a large magnitude for its MMLR. Likewise, the list of ranking genes is made by the marginal strength of association with the response. That is, the top-ranked genes considered as most promising features have larger values of MMLR. To make a decision, the threshold of selecting top genes from the list, SIS would be used. It is a simple and effective algorithm which includes the true significant variables with probability tending toward one [43]. The cutoff value to select top-ranked genes is set up with $\left\lfloor \frac{n}{\log(n)} \right\rfloor$. Those filtered genes would be plugged into the sparse logistic regression models such as LASSO, MCP, and SCAD to further evaluate the performance of classification as well as gene selection. The Algorithm 1 describes the procedure of the proposed two-stage approach.

Algorithm 1 Proposed two-step procedure

Step 1: Sample 70% of samples randomly without replacement from the training set.
Step 2: Count frequency of each of genes from 100 models of λ values.
Step 3: Repeat Step 1 and Step 2 100 times.
Step 4: Calculate selection probability for each of variables based on Equation (10) and then rank them.
Step 5: Select top $\left\lfloor \frac{n}{\log(n)} \right\rfloor$ genes with the highest frequency.
Step 6: Apply them to sparse logistic regression methods to build prognostic models.

2.3. The Proposed Variable Ranking Method

We utilize the following elastic net ($\alpha = 0.5$) penalized regression method based on resampling technique to rank the features of importance using frequency. Elastic net is a combination of L_1(LASSO) and L_2(Ridge) and it has the benefit of performing well with highly correlated variables.

$$\hat{\beta}_{elastic\ net} = \underset{\beta}{\operatorname{argmin}}\left[-\sum_{i=1}^{n}(y_i \log(p(x_i)) + (1-y_i)\log(1-p(x_i))) + \lambda\left(\frac{1-\alpha}{2}\sum_{i=1}^{d}|\beta_i|^2 + \alpha\sum_{i=1}^{d}|\beta_i|\right)\right]. \quad (9)$$

The following is the equation of selection probability in each gene based on the elastic net.

$$SP(g_l) = \frac{1}{K}\sum_{i=1}^{K}\frac{1}{L}\sum_{j=1}^{L}I(\beta_i \neq 0), \text{ for } l = 1,2,\ldots,d, \quad (10)$$

where K is the number of resampling, L is the number of λ, β_i is the regression coefficient corresponding gene l, and $I(\ )$ is the indicator variable. In each of K resampling, 100 values of λ are considered to build variable selection models. With SIS approach, top genes are selected and then applied those genes to LASSO, MCP, and SCAD penalized logistic regression method. The following is the algorithm of our proposed filter ranking method to rank the variable of importance. Figure 1 describes the schema of the proposed two-step approach.

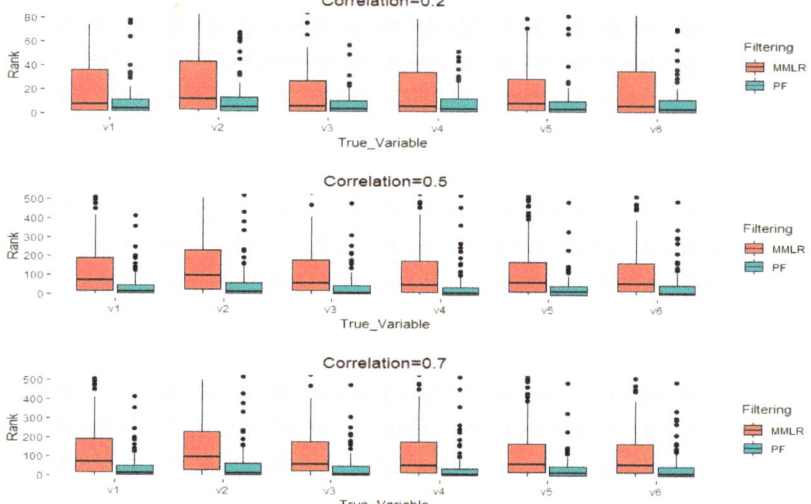

Figure 2. The boxplots of ranking true variables with the proposed filter method (PF) and MMLR method under correlation coefficients 0.2, 0.5, and 0.7 with 100 iterations.

2.4. Metrics of Performance

We calculated accuracy, the geometric mean of sensitivity and specificity (G-mean), and area under the receiver operating characteristic curve (AUROC). The accuracy is done with the following equation:

$$\text{Accuracy} = \frac{TP + TN}{TP + FP + TN + FN} \times 100, \qquad (11)$$

where TP is the number of true positives, TN the number of true negatives, FP the number of false positives, and FN the number of false negatives.

The geometric mean of sensitivity and specificity was used to check the joint performance. The equation is as follows:

$$\text{Geometric mean} = \sqrt{\text{Sensitivity} \times \text{Specificity}}. \qquad (12)$$

AUROC was also considered to evaluate the overall classification performance of the proposed method. A perfect overall classification produces an AUROC = 1 whereas a random overall classification has an AUROC = 0.5.

3. Results

3.1. Simulation Results

The response variable is generated by a sequence of Bernoulli trial with the following probability:

$$\pi_i(y_i = 1 | x_i) = \frac{\exp(x_i \beta)}{1 + \exp(x_i \beta)}. \qquad (13)$$

Data in each iteration are generated by using a multivariate normal distribution with mean 0 and variance-covariance matrix Σ with compound symmetry correlation structure whose diagonal elements are 1 and off-diagonal elements are $\rho = 0.2$, 0.5, and 0.7, respectively. The following is the variance-covariance matrix:

$$\Sigma = \begin{pmatrix} 1 & \rho & \cdots & \rho \\ \rho & 1 & \cdots & \rho \\ \vdots & \vdots & \ddots & \vdots \\ \rho & \rho & \cdots & 1 \end{pmatrix}_{d \times d} \qquad (14)$$

$x_j \sim N_d(0, \Sigma)$ is the jth row of design matrix X and y_i is a binary outcome generated by a Bernoulli trial with the probability from Equation (13). 100 datasets, where n is 200 and d is 1000, are generated and six true regression coefficients are generated from a uniform distribution with min and max values which are 2 and 4, respectively. The simulation data are applied to PF as well as MMLR as a first stage to show the superiority of performance that true variables are highly ranked. The variable ranking procedure in PF was run 100 times with resampling technique. Then the calculated average selection probabilities of each of the 1000 variables were used to rank them. The result of filtering performance was summarized as boxplots described in Figure 2. As seen in boxplots with three different correlation structures, the ranking of six true important variables is higher than that of MMLR. Under the correlation coefficient of 0.2, the average ranking of the six true variables with the proposed ranking method was at 22nd among 1000 variables, whereas the MMLR method was at 44th. In case of high correlation coefficients of 0.5 and 0.7, the proposed one was 59th and 62nd while MMLR was 132nd and 139th.

In addition, an average number of true variables included in filtered data with SIS is reported in Table 1. As seen in Table 1, the proposed method includes more true variables than MMLR in the various correlation settings. For each correlation setting, we used a paired two-sample *t*-test to check for significance level for the mean difference of the true number of variables between the two methods through 100 iterations, and all three were significant. That is, the proposed method is superior to MMLR for filtering true variables with SIS.

Table 2 shows that the performance of prediction as well as geometric mean with SIS-LASSO, SIS-MCP, and SIS-SCAD based on the proposed filter method are better than that of MMLR. As seen in TP (average number of true positives) of Table 2, all three variable selection methods capture mostly a true number of variable filtered from each of PF and MMLR. However, model size (MS) with the proposed filter ranking method is larger than that of MMLR because the methods with more true variables have a tendency to select unimportant variables highly correlated with the true variables. Figure 3 shows the boxplots of the area under the receiver operating characteristic (AUROC) for each of three methods with both proposed filter ranking and MMLR ranking methods based on SIS under three different correlation coefficients ($\rho = 0.2, 0.5,$ and 0.7). It also demonstrated that the AUROCs of SIS-methods based on the proposed filter ranking method is better performed compared to those of MMLR.

Table 1. An average number of true positives from the proposed PF and MMLR with SIS and a significance level of paired two-sample t-test for the mean difference of the number of true positives between two methods using the number of true positives obtained over 100 iterations.

Filtering Method	Metric	Correlation Coefficient		
		0.2	0.5	0.7
PF	Number of True Positive	5.4 (0.765)	4.21 (1.09)	3.11 (1.09)
MMRL		4.52(0.948)	2.15 (1.26)	0.29 (0.50)
two sample t-test (p value)		1.204×10^{-11}	$< 2.2 \times 10^{-16}$	$< 2.2 \times 10^{-16}$

*(): standard deviation.

The variable selection procedures of SIS-LASSO, SIS-MCP, and SIS-SCAD with both PF and MMRL filtered data were run 100 times using compound symmetry correlation structure with 0.2, 0.5, and 0.7. In each iteration, accuracy, area under the receiver operating characteristic (AUROC), geometric mean (G-mean) for sensitivity and specificity, true positives (TP), and false positives (FP). The results of performance for the variable selection methods with both filter ranking methods are summarized in Table 2.

3.2. Real Data Analysis

To test the performance of SIS-LASSO, SIS-MCP, and SIS-SCAD after filtering with the proposed method, we analyzed colon cancer gene expression data. The dataset contains 62 samples, which included 40 colon tumors and 22 normal colon tissue samples and 2000 genes whose gene expression information was extracted from DNA microarray data resulting from preprocessing; all 2,000 genes have unique expressed tags (ESTs) named. We also analyzed lung cancer gene expression data, GSE10072. The dataset includes 107 samples, which are made up of 49 normal lung and 58 lung tumor samples with 22,283 genes. Initially, we calculated the pairwise correlation for the normal and cancer samples combined to check the extent of overall correlation among genes in the colon cancer. The pairwise correlation is summarized in Figure 4 as a histogram with boxplot. The mean correlation between genes is 0.428 with a standard deviation of 0.203. It is clear that there is a high correlation between genes and this falls between the values tested in the simulation studies. In case of the lung cancer, the mean correlation between genes is 0.012 with a standard deviation of 0.246 because we used a full gene expression data unlike the colon gene expression data.

Table 2. Classification performance of proposed filtering (PF) compared to marginal maximum likelihood logistic regression estimator (MMLR) with SIS-LASSO, SIS-MCP, and SIS-SCAD over 100 iterations.

Correlation	Filtering	Methods	Accuracy	G-mean	TP	FP	MS
0.2	PF	SIS-LASSO	0.856(0.047)	0.854(0.049)	5.25(0.757)	0.019(0.002)	24.55(1.971)
		SIS-MCP	0.878(0.054)	0.877(0.056)	5.03(0.937)	0.006(0.003)	11.3(2.805)
		SIS-SCAD	0.878(0.053)	0.876(0.055)	5.18(0.757)	0.012(0.005)	17.24(5.053)
		average	0.871(0.051)	0.869(0.053)	5.153(0.817)	0.012(0.003)	17.697(3.276)
	MMLR	SIS-LASSO	0.847(0.056)	0.844(0.06)	4.3(0.99)	0.015(0.002)	18.73(2.131)
		SIS-MCP	0.86(0.061)	0.858(0.063)	4.21(0.988)	0.006(0.003)	10.32(2.449)
		SIS-SCAD	0.861(0.059)	0.858(0.062)	4.3(0.99)	0.011(0.004)	14.8(3.649)
		average	0.856(0.059)	0.853(0.062)	4.27(0.989)	0.011(0.003)	14.617(2.743)
0.5	PF	SIS-LASSO	0.886(0.041)	0.884(0.042)	3.65(1.266)	0.019(0.003)	22.71(2.267)
		SIS-MCP	0.869(0.055)	0.868(0.057)	2.93(1.409)	0.008(0.003)	10.87(2.058)
		SIS-SCAD	0.884(0.048)	0.883(0.05)	3.57(1.257)	0.017(0.004)	20.06(3.92)
		average	0.88(0.048)	0.878(0.05)	3.383(1.311)	0.015(0.003)	17.88(2.748)
	MMLR	SIS-LASSO	0.865(0.046)	0.863(0.047)	1.84(1.237)	0.015(0.003)	17.02(2.137)
		SIS-MCP	0.858(0.048)	0.857(0.048)	1.66(1.233)	0.008(0.002)	9.89(1.681)
		SIS-SCAD	0.863(0.047)	0.861(0.047)	1.83(1.28)	0.014(0.003)	15.64(2.873)
		average	0.862(0.047)	0.86(0.047)	1.777(1.25)	0.012(0.003)	14.183(2.23)
0.7	PF	SIS-LASSO	0.911(0.037)	0.911(0.038)	2.74(1.16)	0.019(0.003)	21.14(2.274)
		SIS-MCP	0.899(0.042)	0.899(0.043)	1.82(1.158)	0.007(0.002)	8.88(1.981)
		SIS-SCAD	0.907(0.038)	0.907(0.038)	2.68(1.171)	0.016(0.004)	18.88(3.699)
		average	0.906(0.039)	0.906(0.04)	2.413(1.163)	0.014(0.003)	16.3(2.651)
	MMLR	SIS-LASSO	0.887(0.037)	0.886(0.037)	0.26(0.543)	0.014(0.002)	13.72(1.724)
		SIS-MCP	0.881(0.04)	0.88(0.041)	0.21(0.498)	0.008(0.002)	7.75(1.591)
		SIS-SCAD	0.888(0.036)	0.888(0.037)	0.25(0.52)	0.013(0.002)	13.45(2.285)
		average	0.885(0.038)	0.885(0.038)	0.24(0.52)	0.012(0.002)	11.64(1.867)

*(): standard deviation.

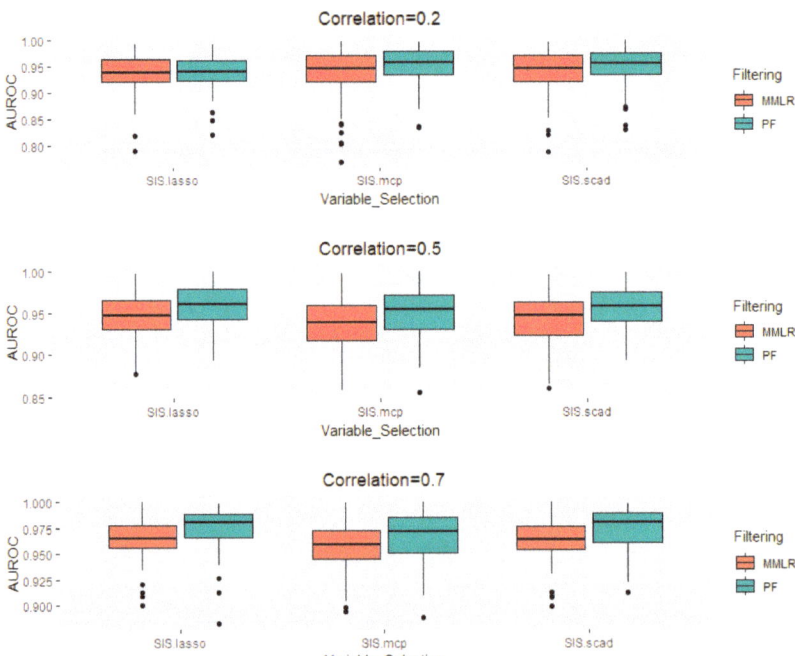

Figure 3. Comparison of area under the receiver operating characteristic (AUROC) with SIS-LASSO, SIS-MCP, and SIS-SCAD after filtering with both proposed filter ranking method and MMLR method under three correlation settings.

To obtain reliable results of the performance of accuracy, AUROC, and G-mean with screened variables, we iterated 100 times of both the colon and lung cancer data with resampling technique. In each iteration, we firstly divided the data into a training set of 70% of samples and a testing set of 30% of samples. Secondly, we select top ranked number of genes with SIS to plug into LASSO, MCP, and SCAD. Finally, we select genes with non-zero coefficients in the model and estimate the performance. We also count genes appeared in the models across three variable selection methods to build lists of ranking genes.

As in the simulation studies, we estimated the average of accuracy, AUROC, G-mean, and model size as the results of using three methods with PF. The results are reported in Table 3. SIS-LASSO with the performance of accuracy and AUROC, each of which is 0.803 and 0.886 with the standard deviations of 0.098 and 0.077 for colon and 0.976 and 0.998 with standard of 0.017 and 0.007, respectively, is relatively better compared to those of other variable selection methods in both datasets. We also presented the top 10 genes selected from each of the three lists of ranking genes across the three variable selection methods based on 100 resampling for the colon cancer and lung cancer data in both Tables 4 and 5. There are eight common genes of G50753, M76378, H08393 H55916, M63391, T62947, R80427, and T71025 among top 10 ranked genes from the results of three methods in the colon data.

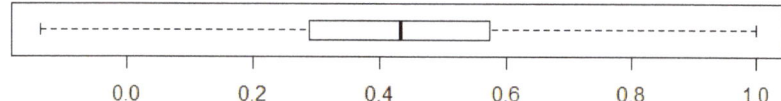

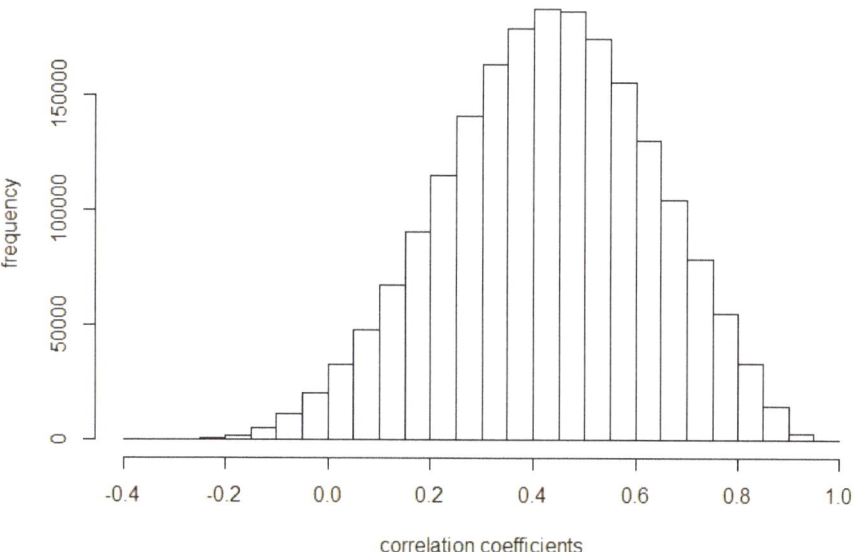

Figure 4. The histogram and boxplot of pairwise correlation coefficients between 2000 expression levels of genes for the colon and normal group combined. The number of correlation coefficients is 1,999,000. Two plots show that average pairwise correlation is 0.428 (median = 0.433) with a standard deviation of 0.203.

The gene of R87126 is common between the results of SIS-LASSO and SIS-MCP, T47377 between SIS-LASSO and SIS-SCAD, and T64012 between SIS-SCAD and SIS-MCP. In particular, G50753, H08393, and H55916 were consistently ranked.

Table 3. Classification performance of the proposed selection method with SIS-LASSO, SIS-MCP, and SIS-SCAD in both colon and lung cancer. It is the average performance resulting from 100 iterations.

Dataset	Method	Accuracy	AUROC	G-Mean	Model Size
	SIS-LASSO	0.803 (0.098)	0.886 (0.077)	0.745 (0.144)	7.8 (1.47)
Colon	SIS-MCP	0.793 (0.097)	0.864 (0.088)	0.748 (0.132)	4.14 (1.054)
	SIS-SCAD	0.798 (0.096)	0.874 (0.082)	0.753 (0.13)	6.73 (1.896)
	SIS-LASSO	0.976 (0.017)	0.998 (0.007)	0.975 (0.019)	9.53 (1.453)
Lung	SIS-MCP	0.952 (0.03)	0.983 (0.017)	0.95 (0.032)	1.09 (0.288)
	SIS-SCAD	0.975 (0.021)	0.997 (0.006)	0.973 (0.023)	8.65 (2.222)

(): standard deviation.

G50753, M63391, and M76378 were reported as significant genes related to colon cancer in [45]. M76378, H08393, H55916, M63391, R87126, and T47377 were also reported as genes associated with colon cancer in [46]. In addition, H08393 (collagen alpha 2(XI) chain) involved in cell adhesion is also known as a gene related to colon carcinoma whose cell has collagen-degrading activity as part of the metastatic process. T62947 has the potential to affect colon cancer by playing a role in controlling cell growth and proliferation through the selective translation of particular classes of mRNA. R80427 is also identified as genes distinguishing colon cancer in [47].

Table 4. Top 10 ranked genes with highest selection frequency from the lists of ranking genes using 100 times resampling approach across three methods of SIS-LASSO, SIS-MCP, and SIS-SCAD on both the colon cancer and the lung cancer gene expression data.

Rank	SIS-LASSO	SIS-MCP	SIS-SCAD
		Gene Accession ID	
1	Hsa.36689 *** (G50753)	Hsa.36689	Hsa.36689
2	Hsa.692.2 *** (M76378)	Hsa.8147	Hsa.692.2
3	Hsa.6814 *** (H08393)	Hsa.6814	Hsa.6814
4	Hsa.1660 *** (H55916)	Hsa.1660	Hsa.1660
5	Hsa.8147 *** (M63391)	Hsa.692.2	Hsa.33268
6	Hsa.5392 *** (T62947)	Hsa.12241 ** (T64012)	Hsa.12241
7	Hsa.37937 ** (R87126)	Hsa.33268	Hsa.5392
8	Hsa.33268 *** (R80427)	Hsa.5392	Hsa.8147
9	Hsa.3016 ** (T47377)	Hsa.8125	Hsa.8125
10	Hsa.8125 *** (T71025)	Hsa.37937	Hsa.3016

***: common genes in all three ranked gene lists, **: common genes in two of the three ranked gene lists.; (): GenBank Accession Number.

Likewise, the top 10 ranked genes in Table 4 from SIS-LASSO, SIS-MCP, and SIS-SCAD with PF were shown to play an important role in colon cancer. Figure 5 shows the boxplots of significantly differentially expressed genes between normal and colon samples on the eight genes found in all three methods. H08393 and H55916 are significantly expressed and downregulated while the other six are upregulated. In case of lung cancer data, there are five common genes of 21957_s_at, 209555_s_at, 209875_s_at, 209074_s_at, and 219213_at among top 10 ranked genes in lung cancer. The genes of 205357_s_at, 203980_at, 208982_at, and 220,170 are common between the results of SIS-LASSO and

SIS-MCP. The gene of 32625_at is common gene between SIS-LASSO and SIS-SCAD. Specially, first top four genes between the results of SIS-LASSO and SIS-SCAD have the same ranking. In addition, there are four unique genes of 209614_at from SIS-SCAD, 206209_s_at, 204271_s_at, 204396_s_at, and 219719_at from SIS-MCP. 219597_s_at (DUOX1) usually is downregulated and associated with lung breast cancer [48,49]. 209555_s_at (CD36) is also related to breast cancer [50] and affects the progression of lung cancer [51]. 209875_s_at (SPP1) is reported as a prognostic biomarker for lung adenocarcinoma [52,53]. 209074_s_at (FAM107A) is also emphasized as a lung cancer biomarker downregulated [54]. Although 219213_at (JAM2) are not directly known as a variant of lung cancer, it is worthwhile to be further investigated as a potential biomarker related to lung adenocarcinoma. We also found that most of five common genes play significant roles in lung cancer. Figure 6 also represents the boxplots of significantly differentially expressed genes between normal and colon samples on the five genes found commonly in the top ten ranked genes in all three methods. Only the gene of 209875_s_at (SPP1) is upregulated while the rest of them are downregulated.

Table 5. Top 10 ranked genes with highest selection frequency from lists of gene ranking using 100 times resampling approach of three methods of SIS-LASSO, SIS-MCP, and SIS-SCAD on the lung cancer gene expression data.

Rank	SIS-LASSO	SIS-MCP	SIS-SCAD
-		Gene Accession ID	
1	219597_s_at ***(DUOX1)	209555_s_at	219597_s_at
2	205357_s_at **	209074_s_at	205357_s_at
3	209555_s_at ***(CD36)	32625_at	209555_s_at
4	209875_s_at ***(SPP1)	206209_s_at *	209875_s_at
5	203980_at **	204271_s_at *	209074_s_at
6	208982_at **	204396_s_at *	219213_at
7	209074_s_at *** (FAM107A)	219213_at	208982_at
8	220170_at **	219597_s_at	220170_at
9	219213_at *** (JAM2)	219719_at *	209614_at *
10	32625_at **	209875_s_at	203980_at

***: common genes in all three ranked gene lists, **: common genes in two of the three ranked gene lists. *: unique genes. (): Gene symbol.

4. Discussion

We explored the feasibility of using the proposed feature ranking method as a filtering stage with Elastic net ($\alpha = 0.5$) based on a resampling approach followed by SIS as screening in conjunction with LASSO, MCP, and SCAD penalized logistic variable selection methods in high dimensional settings to improve the performance of variable selection and classification prediction. One of the currently popular methods achieving such a goal is MMLR. It ranks variables in order from largest to smallest scores of maximum likelihood. It performs poorly with important variables that are marginally weak but jointly and strongly associated with the response since it screens out such variables. The simulation studies demonstrated that the PF method retained more true important variables when compared to MMLR in Table 1. PF method also showed a better performance of retaining a true number of variables as the correlation of the variables was increased than MMRL. It is clear that the elastic net-based PF takes into account correlation among true important variables, while MMLR only considers marginal strength with the outcome variable. However, as seen in the results of using three variable selection methods with SIS based on both filtered data, the proportion of unimportant variables in the models is still high.

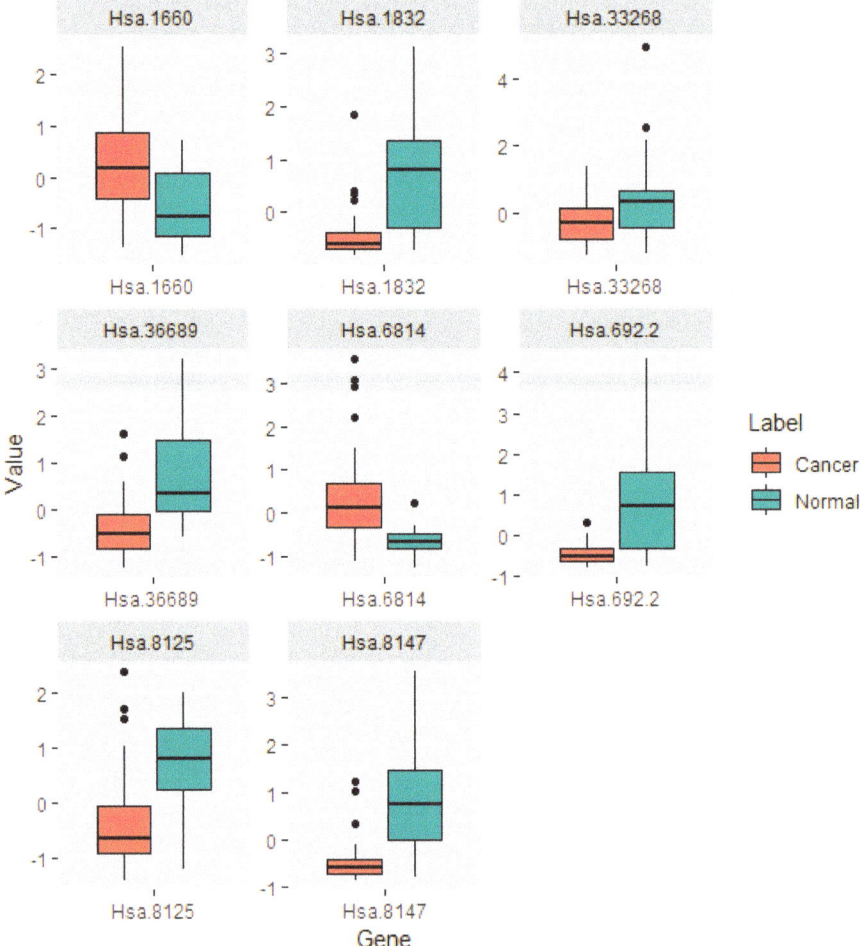

Figure 5. Boxplots of differential expression level between normal and colon samples on eight genes from SIS-LASSO, SIS-MCP, and SIS-SCAD with the ranked data. Each boxplot contains the *p*-value of mean differential expression between two groups with a two-sample *t*-test.

For further confirmation of the PF in selecting the most promising genes for superior classification performance, we applied it to a real example of both colon and lung cancer gene expression data. The SIS-LASSO method produced the best performance scores compared to SIS-MCP and SIS-SCAD. We also selected the top 10 ranked genes with highest selection frequency from the lists of ranking genes generated by the resampling approach in each of three variable selection methods to check gene selection consistency as well as biological significance connecting to colon and lung cancer. There were eight and five overlapped genes among top 10 ranked genes from the results of three methods in Tables 4 and 5, respectively. Most of the genes are reported as significant genes related to colon and lung carcinoma. In addition, some of the genes was consistently highly ranked across the three methods.

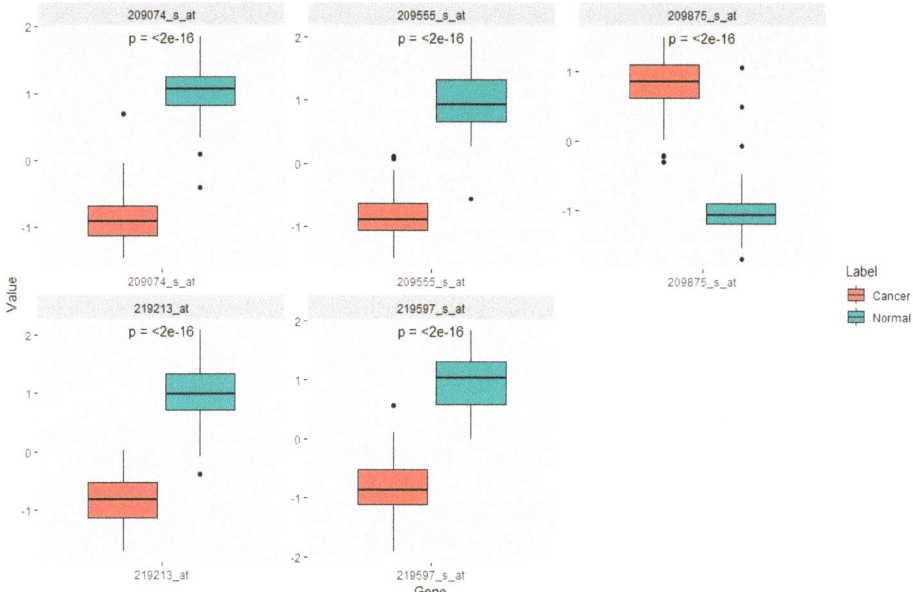

Figure 6. Boxplots of differential expression level between normal and lung samples on five genes from SIS-LASSO, SIS-MCP, and SIS-SCAD with the ranked data. Each boxplot contains the *p*-value of mean differential expression between two groups with a two-sample *t*-test.

5. Conclusions

In this study, the proposed PF demonstrated the superiority of ranking true variables highly as a filtering stage compared to MMLR through extensive simulation studies. Furthermore, the combination of SIS-LASSO, SIS-MCP, and SIS-SCAD with the PF also had better performance of classification as well as detection of true important variables than those with MMLR. Even in real applications of colon and lung gene expression data, it was demonstrated that the proposed two-stage procedure with PF consistently captures the most promising features related to colon and lung cancer. As future research, we plan to develop the methodology of variable selection with PF to increase the power of detecting true important variables as well as prediction of classification.

Author Contributions: All authors drafted the manuscript, and read and approved the final manuscript.

Funding: This research received no external funding.

Acknowledgments: We thank the referees for reading the manuscript very carefully and making a number of valuable and kind comments that improve the presentation of the manuscript.

Conflicts of Interest: The authors declare no conflict of interest.

References

1. Sangjin, K.; Susan, H. High Dimensional Variable Selection with Error Control. *Biomed. Res. Int. Vol.* **2016**, *2016*. [CrossRef]
2. Shuangge, M.; Jian, H. Penalized feature selection and classification in bioinformatics. *Brief. Bioinform.* **2008**, *9*, 392–403.
3. Abhishek, B.; Shailendra, S. Gene Selection Using High Dimensional Gene Expression Data: An Appraisal. *Curr. Bioinform.* **2018**, *13*, 225–233. [CrossRef]

4. Hassan, T.; Elf, E.; lan, W. An efficient approach for feature construction of high-dimensional microarray data by random projections. *PLoS ONE* **2018**, *13*, e0196385. [CrossRef]
5. Bourgon, R. Independent filtering increases detection power for high-throughput experiments. *Proc. Natlacad. Sci.* **2010**, *107*, 9546–9951. [CrossRef] [PubMed]
6. Bourgon, R.; Gentleman, R.; Huber, W. Reply to Talloen et al.: Independent filtering is a generic approach that needs domain-specific adaptation. *Proc. Natl Acad. Sci. USA* **2010**, *107*, E175. [CrossRef]
7. Lu, J.; Peddada, S.D.; Bushel, P.R. Principal component analysis-based filtering improves detection for Affymetrix gene expression arrays. *Nucleic Acids Res.* **2011**, *e86*, 39. [CrossRef]
8. Jiang, H.; Doerge, R.W. A two-step multiple comparison procedure for a large number of tests and multiple treatments. *Stat. Appl. Genet. Mol. Biol.* **2006**, *5*. [CrossRef]
9. Ramskold, E.; Kerns, R.T. An abundance of ubiquitously expressed genes revealed by tissue transcriptome sequence data. *PLoS Comput. Biol.* **2009**, *5*, e1000598. [CrossRef]
10. Sultan, M.; Schulz, M.H.; Richard, H.; Magen, A.; Klingenhoff, A.; Scherf, M.; Seifert, M.; Borodina, T.; Soldatov, A.; Parkhomchuk, D.; et al. A global view of gene activity and alternative splicing by deep sequencing of the human transcriptome. *Science* **2008**, *321*, 956–960. [CrossRef]
11. Calle, M.L.; Urrea, V.; Malats, V.N.; Steen, K.V. Improving strategies for detecting genetic patterns of disease susceptibility in association studies. *Stat. Med.* **2008**, *27*, 6532–6546. [CrossRef] [PubMed]
12. Li, L.; Kabesch, M.; Bouzigon, E.; Demenais, F.; Farrall, M.; Moffatt, M.F.; Lin, X.; Liang, L. Using eQTL weights to improve power for genome-wide association studies: A genetic study of childhood asthma. *Fron. Genet.* **2013**, *4*, 103. [CrossRef] [PubMed]
13. Taqwa, A.A.; Siraj, M.M.; Zainal, A.; Elshoush, H.T.; Elhaj, F. Feature Selection Using Information Gain for Improved Structural-Based Alert Correlation. *PLoS ONE* **2016**, *11*, e0166017. [CrossRef]
14. Tan, Y.; Liu, Z. Feature selection and prediction with a Markov blanket structure learning algorithm. *BMC Bioinform.* **2013**, *14*, A3. [CrossRef]
15. Kakourou, A.; Mertens, B. Bayesian variable selection logistic regression with paired proteomic measurements. *Biom. J.* **2018**. [CrossRef] [PubMed]
16. Kursa, M.B.; Rudnicki, W.R. Feature Selection with the Boruta Package. *J. Stat. Softw.* **2010**, *36*, 1–13. [CrossRef]
17. Okeh, U.M.; Oyeka, I.C.A. Estimating the fisher's scoring matrix formula from the logistic model. *Am. J. Theor. Appl. Stat.* **2013**, *2*, 221–227.
18. Urbanowicz, R.J.; Meekerb, M.; La Cavaa, W.; Olsona, R.S.; Moorea, J.H. Relief-based feature selection: Introduction and review. *J. Biomed. Inform.* **2018**, *85*, 189–203. [CrossRef]
19. Milos, R.; Mohamed, G.; Nenad, F.; Zoran, O. Minimum redundancy maximum relevance feature selection approach for temporal gene expression data. *BMC Bioinform. BMC Ser.* **2017**, *18*, 9. [CrossRef]
20. Algamal, Z.Y.; Lee, M.H. A two-stage sparse logistic regression for optimal gene selection in high-dimensional microarray data classification. *Adv. Data Anal. Classif.* **2018**, 1–19. [CrossRef]
21. Le, T.T.; Urbanowicz, R.J.; Moore, J.H.; McKinney, B.A. Statistical Inference Relief (STIR) feature selection. *Bioinformatics* **2018**, *788*. [CrossRef] [PubMed]
22. Abdel-Aal, R.E. GMDH-based feature ranking and selection for improved classification of medical data. *J. Biomed. Inf.* **2005**, *38*, 456–468. [CrossRef] [PubMed]
23. Fan, J. Sure Independence screening for ultrahigh dimensional feature space. *J. R. Stat. Soc. B* **2008**, *70*, 849–911. [CrossRef]
24. Dizler, G.; Morrison, J.C.; Lan, Y.; Rosen, G.L. Fizzy: Feature subset selection for metagenomics. *BMC Bioinform.* **2015**, *1*, 358. [CrossRef]
25. Peng, H.; Long, F.; Ding, C. Feature selection based on mutual information: Criteria of max-dependency, max-relevance, and min-redundancy. *IEEE Trans. Pattern Anal. Mach. Intell.* **2005**, *27*, 1226–1238. [CrossRef]
26. Wei, M.; Chow, T.W.S.; Chan, R.H.M. Heterogeneous feature subset selection using mutual information based feature transformation. *Neurocomputing* **2015**, *168*, 706–718. [CrossRef]
27. Su, C.-T.; Yang, C.-H. Feature selection for the SVM: An application to hypertension diagnosis. *Expert Syst. Appl.* **2008**, *34*, 754–763. [CrossRef]

28. Tibshirani, R. Regression Shrinkage and Selection via the Lasso. *J. R. Stat. Soc. Ser. B* **1996**, *58*, 267–288. [CrossRef]
29. Zhang, C.-H. Nearly unbiased variable selection under minimax concave penalty. *Ann. Stat.* **2010**, *38*, 894–942. [CrossRef]
30. Fan, J.; Li, R. Variable Selection via Nonconcave Penalized Likelihood and its Oracle Properties. *J. Am. Stat. Assoc.* **2001**, *96*, 1348–1360. [CrossRef]
31. Two-Stage-Resources-2019. Available online: https://sites.google.com/site/sangjinkim0716/data-repository/two-stage-resources-2019 (accessed on 29 May 2019).
32. Pappua, V.; Panagopoulosb, O.P.; Xanthopoulosb, P.; Pardalosa, P.M. Sparse proximal support vector machines for features selection in high dimensional datasets. *Expert Syst. Appl.* **2015**, *42*, 9183–9191. [CrossRef]
33. Liao, J.G.; Chin, K.-V. Logistic regression for disease classification using micro data: Model selection in a large p and small n case. *Bioinformatics* **2007**, *23*, 1945–1951. [CrossRef] [PubMed]
34. Park, M.Y.; Hastie, T. Penalized logistic regression for detecting gene interactions. *Biostatistics* **2008**, *9*, 30–50. [CrossRef] [PubMed]
35. Bielza, C.; Robles, V.; Larrañaga, P. Regularized logistic regression without a penalty term: An application to cancer classification with microarray data. *Expert Syst. Appl.* **2011**, *38*, 5110–5118. [CrossRef]
36. Bootkrajang, J.; Kabán, A. Classification of mislabelled microarrays using robust sparse logistic regression. *Bioinformatics* **2013**, *29*, 870–877. [CrossRef] [PubMed]
37. Cawley, G.C.; Talbot, N.L.C. Gene selection in cancer classification using sparse logistic regression with Bayesian regularization. *Bioinformatics* **2006**, *22*, 2348–2355. [CrossRef] [PubMed]
38. Li, J.; Jia, Y.; Zhao, Z. Partly adaptive elastic net and its application to microarray classification. *Neural Comput. Appl.* **2012**, *22*, 1193–1200. [CrossRef]
39. Sun, H.; Wang, S. Penalized logistic regression for high-dimensional DNA methylation data with case-control studies. *Bioinformatics* **2012**, *28*, 1368–1375. [CrossRef] [PubMed]
40. Zhu, J.; Hastie, T. Classification of gene microarrays by penalized logistic regression. *Biostatistics* **2004**, *5*, 427–443. [CrossRef]
41. Liang, Y.; Liu, C.; Luan, X.-Z.; Leung, K.-S.; Chan, T.-M.; Xu, Z.-B.; Zhang, H. Sparse logistic regression with an L1/2 penalty for gene selection in cancer classification. *BMC Bioinform.* **2013**, *14*, 198–211. [CrossRef]
42. Huang, H.H.; Liu, X.Y.; Liang, Y. Feature selection and cancer classification via sparse logistic regression with the hybrid L1/2 + 2 regularization. *PLoS ONE* **2016**, *11*, e0149675. [CrossRef] [PubMed]
43. Algamal, Z.Y.; Lee, M.H. Penalized logistic regression with the adaptive LASSO for gene selection in high-dimensional cancer classification. *Expert Syst. Appl.* **2015**, *42*, 9326–9332. [CrossRef]
44. Ben Brahim, A.; Limam, M. A hybrid feature selection method based on instance learning and cooperative subset search. *Pattern Recogn. Lett.* **2016**, *69*, 28–34. [CrossRef]
45. Wang, Y.; Yang, X.-G.; Lu, Y. Informative Gene Selection for Microarray Classification via Adaptive Elastic Net with Conditional Mutual Information. *Appl. Math. Model.* **2019**, *71*, 286–297. [CrossRef]
46. Patrick, M.; John, S.; Rebecca, W. Methods for Bayesian Variable Selection with Binary Response Data using the EM algorithm. *arXiv* **2016**, arXiv:1605.05429.
47. Castellanos-Garzon, J.A.; Ramos-Gonzalez, J. A Gene Selection Approach based on Clustering for Classification Tasks in Colon Cancer. *Adv. Distrib. Comput. Artif. Intell. J.* **2015**, *4*. [CrossRef]
48. Fortunato, R.S.; Gomes, L.R.; Munford, V.; Pessoa, C.F.; Quinet, A.; Hecht, F.; Kajitani, G.S.; Milito, C.B.; Carvalho, D.P.; Martins Menck, C.F. DUOX1 Silencing in Mammary Cell Alters the Response to Genotoxic Stress. *Oxid. Med. Cell. Longev.* **2018**, *2018*. [CrossRef] [PubMed]
49. Little, A.C.; Sham, D.; Hristova, M.; Danyal, K.; Heppner, D.E.; Bauer, R.A.; Sipsey, L.M.; Habibovic, A.; van der Vliet, A. DUOX1 silencing in lung cancer promotes EMT, cancer stem cell characteristics and invasive properties. *Oncogenesis* **2016**, *5*. [CrossRef] [PubMed]
50. Liang, Y.; Han, H.; Liu, L.; Duan, Y.; Yang, X.; Ma, C.; Zhu, Y.; Han, J.; Li, X.; Chen, Y. CD36 plays a critical role in proliferation, migration and tamoxifen-inhibited growth of ER-positive breast cancer cells. *Oncogenesis* **2018**, *7*, 98. [CrossRef] [PubMed]

51. Sun, Q.; Zhang, W.; Guo, F. Hypermethylated CD36 gene affected the progression of lung cancer. *Genetics* **2018**, *678*, 395–406. [CrossRef] [PubMed]
52. Zhang, W.; Fan, J.; Chen, Q.; Lei, C.; Qiao, B.; Liu, Q. SPP1 and AGER as potential prognostic biomarkers for lung adenocarcinoma. *Oncol. Lett.* **2018**, *15*, 7028–7036. [CrossRef] [PubMed]
53. Ioanna, G.; Vasilieios, P.; Ioannis, L.; Nikolaos, K.; Theodora, A.; Georgios, S. Tumor cell-derived osteopontin promotes lung metastasis via both cell-autonomous and paracrine pathways. *Eur. Respir. J.* **2016**, *48*. [CrossRef]
54. Pastuszak-Lewandoska, D.; Czarnecka, K.H.; Nawrot, E.; Domanska, D.; Kiszalkiewicz, J. Decreased FAM107A Expression in Patients with Non-small Cell Lung Cancer. *Adv. Exp. Med. Biol.* **2015**, *852*, 39–48. [PubMed]

© 2019 by the authors. Licensee MDPI, Basel, Switzerland. This article is an open access article distributed under the terms and conditions of the Creative Commons Attribution (CC BY) license (http://creativecommons.org/licenses/by/4.0/).

Article

An Estimation of Sensitive Attribute Applying Geometric Distribution under Probability Proportional to Size Sampling

Gi-Sung Lee [1], Ki-Hak Hong [2] and Chang-Kyoon Son [3,*]

1. Department of Children Welfare, Woosuk University, Wanju Jeonbuk 55338, Korea; gisung@woosuk.ac.kr
2. Department of Computer Science, Dongshin University, Naju Jeonnam 58245, Korea; khhong@dsu.ac.kr
3. Department of Applied Statistics, Dongguk University, Gyeongju Gyeongbuk 38066, Korea
* Correspondence: sonchangkyoon@gmail.com

Received: 29 August 2019; Accepted: 31 October 2019; Published: 14 November 2019

Abstract: In this paper, we extended Yennum et al.'s model, in which geometric distribution is used as a randomization device for a population that consists of different-sized clusters, and clusters are obtained by probability proportional to size (PPS) sampling. Estimators of a sensitive parameter, their variances, and their variance estimators are derived under PPS sampling and equal probability two-stage sampling, respectively. We also applied these sampling schemes to Yennum et al.'s generalized model. Numerical studies were carried out to compare the efficiencies of the proposed sampling methods for each case of Yennum et al.'s model and Yennum et al.'s generalized model.

Keywords: probability proportional to size (PPS) sampling; geometric distribution; sensitive attribute; randomization device; Yennum et al.'s model

1. Introduction

The randomized response model (RRM) was suggested by [1] to estimate the true population proportion of sensitive characteristics, such as illegal gambling, drug-abuse, tax evasion, the extent of illegal income, and the experience of abortion, among others [2–4].

Since Warner's work, many scholars have developed the RRM in various ways. In [5,6], they arranged, summarized, and systemized various RRMs and emphasized their importance. In [7], sampling survey of sensitive attributes applied two-stage cluster sampling to RRM for a population consisting of equal-sized clusters, and [8] considered the cluster RRM for a population consisting of different-sized clusters, where the clusters are selected by probability proportional to size (PPS) sampling.

Recently, Yennum et al. [9] suggested a new randomization device to gather sensitive data in two-stages under the assumption of geometric distribution and made a generalization of their model encompassing generalized geometric distribution using [10] model.

Based on Yennum et al.'s work, it is assumed that the respondents are selected by simple random sampling with replacements, but a real survey selects respondents from various sampling schemes.

Now, we can consider a large sample of clusters. For example, to estimate the true population proportion of drug-abuse among high school students, it is possible to use a randomization device like Yennum et al.'s model via proportional sampling by considering the primary sampling unit as the school and the secondary sampling unit as the students.

From this point of view, we extend Yennum et al.'s model, in which geometric distribution is used as a randomization device based on a population that consists of different-sized clusters, and the clusters are selected by PPS sampling. Estimators of a sensitive parameter, their variances, and their variance estimators are derived by PPS sampling and equal probability two-stage sampling, respectively.

We also apply these methods to the case of Yennum et al.'s generalized model. Numerical studies are carried out to compare the efficiencies of the suggested methods in each case of Yennum et al.'s model and Yennum et al.'s generalized model.

2. An Estimation of Sensitive Attributes with Probability Proportional to Size Sampling under Yennum et al.'s Model

In Section 2, we consider a new sampling scheme to estimate sensitive attributes using Yennum et al.'s model, in which geometric distribution is used as a randomization device when n clusters are selected with proportional to size (PPS) sampling or equal probability sampling from a population that consists of N clusters of size, $M_i(i = 1, 2, \cdots, N)$ and $m_i(i = 1, 2, \cdots, n)$ units are selected by simple random sampling from each sampled cluster.

In Section 2.1, we consider the sampling method for the clusters via PPS sampling with replacements. Clusters by PPS sampling without replacement are considered in Section 2.2, and clusters by equal probability sampling are examined in Section 2.3.

2.1. PPS Sampling with Replacement

Let the population be composed of N clusters. In the first stage, the size of the n sample of the first sampling units (FSU) is selected with replacement by the selection probability p_i for the ith cluster. In the second stage, m_i second sampling units (SSU) are drawn by simple random sampling with replacement (SRSWR) from each FSU and are guided to carry out Yennum et al.'s randomization device.

First of all, the randomization device consists of two elements. The first randomization device for the ith cluster consists of two kinds of urns with white and black balls, where the selection probability of a white ball is W_i, and the selection probability of a black ball is $1 - W_i$.

On the other hand, the second randomization device is composed of two kinds of urns with balls. The first device with balls contains a slip of paper including two statements, such as "I have a sensitive attribute" with selection probability P_i, and the other balls includes a statement such as "I do not have a sensitive attribute" with selection probability $1 - P_i$. The second device with balls contains a slip of paper with the statement "I do not have a sensitive attribute" with selection probability T_i and balls with the statement "I have a sensitive attribute" with selection probability $1 - T_i$.

In the first stage, for the ith cluster, each interviewee draws a ball from the first randomization device, such as the urn with the white and black balls. When he or she selects a white ball, he or she is guided to pick balls from the first urn of the second randomization device, one after another, with replacement, until the first ball containing a statement matching his or her own status appears.

We assume that X_{i1} is the total number of balls drawn before he or she obtains the first ball including his or her own status in the ith cluster, and X_{i2} is the total number of balls drawn before he or she obtains the first ball with his or her own status of not having a sensitive attribute in the ith cluster. Similarly, when he or she draws a black ball, he or she is guided to pick balls from the second urn of the second randomization device, one after another, with replacement, until the first ball containing a statement matching his or her own status appears.

For the ith cluster, using the randomization device in Figure 1, the total number of balls taken by interviewees $X_{i1}, X_{i2}, Y_{i1}, Y_{i2}$ are distributed via generalized geometric distribution. Let π_i and $1 - \pi_i$ be the true population proportion of persons who have a sensitive attribute A_i and A_i^c for the ith cluster. Assume that each interviewee in the ith cluster is drawn by SRSWR.

For the ith cluster, the total number for each ball selected by interviewees through the proposed two-stage device distributes one of the following random variables: $X_{i1} \sim Ge(P_i)$, $X_{i2} \sim Ge(1 - P_i)$, $Y_{i1} \sim Ge(T_i)$ and $Y_{i2} \sim Ge(1 - T_i)$, where $Ge(\cdot)$ represents the geometric distribution with a success probability. Let π_i and $1 - \pi_i$ be the true population proportions of persons who have a sensitive attribute (A_i and A_i^c, respectively) for the ith cluster. Assume that each interviewee in the ith cluster is drawn by SRSWR.

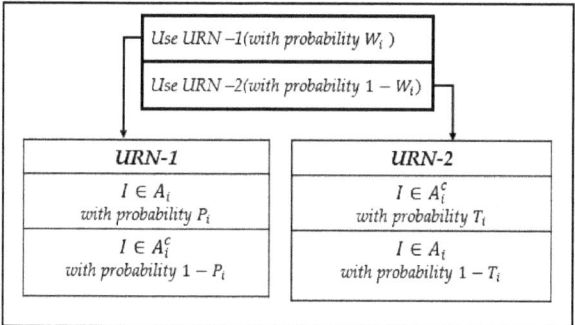

Figure 1. Randomization device for the ith cluster.

Let Z_{ij} be the jth observed answer in the ith cluster; Z_{ij} can be expressed as

$$Z_{ij} = \begin{cases} X_{i1}, \text{ with probability } W_i \pi_i \\ Y_{i2}, \text{ with probability } (1-W_i)\pi_i \\ X_{i2}, \text{ with probability } W_i(1-\pi_i) \\ Y_{i1}, \text{ with probability } (1-W_i)(1-\pi_i) \end{cases} \quad (1)$$

Then, we can find the expected value of Z_{ij} as follows:

$$\begin{aligned} E(Z_{ij}) &= \pi_i \left[\frac{W_i}{P_i} + \frac{(1-W_i)}{(1-T_i)} \right] + (1-\pi_i) \left[\frac{W_i}{(1-P_i)} + \frac{(1-W_i)}{T_i} \right] \\ &= \pi_i \left[\frac{W_i}{P_i} + \frac{(1-W_i)}{(1-T_i)} - \frac{W_i}{(1-P_i)} - \frac{(1-W_i)}{T_i} \right] + \frac{W_i}{(1-P_i)} + \frac{(1-W_i)}{T_i}. \end{aligned} \quad (2)$$

The expected value (2) can be expressed as follows:

$$\frac{(1-T_i)P_i\{E(Z_{ij})(1-P_i)T_i - W_iT_i - (1-W_i)(1-P_i)\}}{P_iT_i(1-P_i)(1-T_i)} = \frac{\pi_i \psi_i}{P_iT_i(1-P_i)(1-T_i)}, \quad (3)$$

where $\psi_i = W_i(1-2P_i)T_i(1-T_i) + (1-W_i)(2T_i-1)P_i(1-P_i)$.

Now the estimator $\hat{\pi}_i$ for the true population proportion π_i in the ith cluster is given by:

$$\hat{\pi}_i = \frac{1}{\psi_i} \left[P_iT_i(1-P_i)(1-T_i)\frac{1}{m_i}\sum_{i=1}^{m_i} Z_{ij} - W_iT_iP_i(1-T_i) - P_i(1-W_i)(1-P_i)(1-T_i) \right]. \quad (4)$$

When the interviewees are drawn by SRSWR from the ith cluster selected with a replacement by the sampling probability p_i, the estimator $\hat{\pi}_{ppswr}$ of the true population proportion π for a sensitive character is given by:

$$\begin{aligned} \hat{\pi}_{ppswr} &= \frac{1}{nM_0} \sum_{i=1}^{n} \frac{M_i \hat{\pi}_i}{p_i} \\ &= \frac{1}{nM_0} \sum_{i=1}^{n} \frac{M_i}{p_i \psi_i} \left[P_iT_i(1-P_i)(1-T_i)\frac{1}{m_i} \sum_{j=1}^{m_i} Z_{ij} - W_iT_iP_i(1-T_i) - P_i(1-W_i)(1-P_i)(1-T_i) \right], \end{aligned} \quad (5)$$

where $M_0 = \sum_{i=1}^{N} M_i$.

Theorem 1: *The estimator $\hat{\pi}_{ppswr}$ of the true population proportion of a sensitive attribute π under PPS with a replacement sampling scheme is an unbiased estimator.*

Proof:

$$E_1 E_2(\hat{\pi}_{ppswr}) = E_1 E_2 \left[\frac{1}{nM_0} \sum_{i=1}^{n} \frac{M_i \hat{\pi}_i}{p_i} \right]$$

$$= E_1 \left[\frac{1}{nM_0} \sum_{i=1}^{n} \frac{M_i E_2(\hat{\pi}_i)}{p_i} \right],$$

and since:

$$E_2(\hat{\pi}_i) = \frac{1}{\psi_i} \left[P_i T_i (1-P_i)(1-T_i) \frac{1}{m_i} \sum_{i=1}^{m_i} E_2(Z_{ij}) - W_i T_i P_i (1-T_i) - P_i (1-W_i)(1-P_i)(1-T_i) \right]$$

$$= \pi_i.$$

we can obtain:

$$E_1 E_2(\hat{\pi}_{ppswr}) = E_1 \left[\frac{1}{nM_0} \sum_{i=1}^{n} \frac{M_i \pi_i}{p_i} \right]$$

$$= \frac{1}{M_0} \sum_{i=1}^{N} p_i \frac{M_i \pi_i}{p_i}$$

$$= \pi.$$

□

Theorem 2: *The variance of $\hat{\pi}_{ppswr}$ is obtained from a two-stage procedure, such that a sample of size n FSU is selected by replacement with sampling probability p_i for the unit i from the population of N clusters with size M_i elements in the ith cluster, and the SSUs with size m_i are drawn by SRSWR from each FSU, as given by:*

$$V(\hat{\pi}_{ppswr}) = \frac{1}{nM_0^2} \sum_{i=1}^{N} p_i \left[\frac{M_i \pi_i}{p_i} - M_0 \pi \right]^2 \quad (6)$$

$$+ \frac{1}{nM_0^2} \sum_{i=1}^{N} \frac{M_i^2}{m_i p_i} \left[\pi_i (1-\pi_i) - \frac{\pi_i}{\psi_i^2} A_i + \frac{1}{\psi_i^2} B_i \right],$$

where:

$$A_i = W_i(1-2P_i)(2-P_i+P_i^2)T_i^2(1-T_i)^2 + (1-W_i)(2T_i-1)(2-T_i+T_i^2)P_i^2(1-P_i)^2$$
$$+ W_i^2 T_i^2 (1-T_i)^2 (2P_i-1) + (1-W_i)^2 P_i^2 (1-P_i)^2 (1-2T_i)$$
$$+ 2W_i(1-W_i) P_i T_i (1-P_i)(1-T_i)(P_i-T_i),$$

$$B_i = (1-T_i)^2 P_i^2 \{ W_i(1-W_i)(P_i+T_i-1)^2 + W_i P_i T_i^2 + (1-W_i)(1-T_i)(1-P_i)^2 \}.$$

Proof: Given $X_{i1} \sim G(P_i)$, $X_{i2} \sim G(1-P_i)$, $Y_{i1} \sim G(T_i)$, $Y_{i2} \sim G(1-T_i)$, where G represents the geometric distribution with a success probability. Since the expected values of Z_{ij} and Z_{ij}^2 are

$$E(Z_{ij}) = \pi_i \left[\frac{W_i}{P_i} + \frac{(1-W_i)}{(1-T_i)} \right] + (1-\pi_i) \left[\frac{W_i}{(1-P_i)} + \frac{(1-W_i)}{T_i} \right], \quad (7)$$

then:

$$E(Z_{ij}^2) = \pi_i \left[\frac{W_i(2-P_i)}{P_i^2} \right] + (1-\pi_i) \left[\frac{W_i(1+P_i)}{(1-P_i)^2} \right] + (1-\pi_i)(1-V_i) \left[\frac{(2-T_i)}{T_i^2} \right] + \pi_i (1-W_i) \left[\frac{(1+T_i)}{(1-T_i)^2} \right]. \quad (8)$$

Based on (7) and (8), the variance of Z_{ij} is:

$$\sigma_{iZ}^2 = E(Z_{ij}^2) - \left[E(Z_{ij}) \right]^2, \quad (9)$$

and, since Z_{ij} is independent, the variance of $\hat{\pi}_i$ can be expressed by:

$$V(\hat{\pi}_i) = V\left[\frac{1}{\psi_i}\left(\frac{P_iT_i(1-P_i)(1-T_i)}{m_i}\sum_{j=1}^{m_i} Z_{ij} - W_iT_iP_i(1-T_i) - P_i(1-W_i)(1-P_i)(1-T_i)\right)\right]$$
$$= \frac{P_i^2T_i^2(1-P_i)^2(1-T_i)^2}{n_h^2\psi_h^2}\sum_{j=1}^{m_i}\sigma_{iZ}^2 \quad (10)$$
$$= \frac{\pi_i(1-\pi_i)}{m_i} + \frac{\pi_i}{m_i\psi_i^2}A_i + \frac{1}{m_i\psi_i^2}B_i.$$

Since $V(\hat{\pi}_{ppswr}) = V_1E_2(\hat{\pi}_{ppswr}) + E_1V_2(\hat{\pi}_{ppswr})$, then the first and second terms are given, respectively, as:

$$V_1E_2(\hat{\pi}_{ppswr}) = V_1E_2\left[\frac{1}{nM_0}\sum_{i=1}^{n}\frac{M_i\hat{\pi}_i}{p_i}\right]$$
$$= V_1\left[\frac{1}{nM_0}\sum_{i=1}^{n}\frac{M_i\pi_i}{p_i}\right]$$
$$= \frac{1}{nM_0^2}\sum_{i=1}^{N}p_i\left[\frac{M_i\pi_i}{p_i} - M_0\pi\right]^2,$$

and

$$E_1V_2(\hat{\pi}_{ppswr}) = E_1V_2\left[\frac{1}{nM_0}\sum_{i=1}^{n}\frac{M_i\hat{\pi}_i}{p_i}\right]$$
$$= E_1\left[\frac{1}{(nM_0)^2}\sum_{i=1}^{n}\frac{M_i^2}{p_i^2}V_2(\hat{\pi}_i)\right]$$
$$= E_1\left[\frac{1}{(nM_0)^2}\sum_{i=1}^{n}\frac{M_i^2}{p_i^2}V_2\left\{\frac{1}{\psi_i}P_iT_i(1-P_i)(1-T_i)\frac{1}{m_i}\sum_{i=1}^{m_i}Z_{ij} - W_iT_iP_i(1-T_i) - P_i(1-W_i)(1-P_i)(1-T_i)\right\}\right]$$
$$= E_1\left[\frac{1}{(nM_0)^2}\sum\frac{M_i^2}{m_ip_i^2}\left\{\pi_i(1-\pi_i) + \frac{\pi_i}{\psi_i^2}A_i + \frac{1}{\psi_i^2}B_i\right\}\right].$$

Then, we can obtain the variance (10).

Moreover, an unbiased estimator of $V(\hat{\pi}_{ppswr})$ is given by

$$\hat{V}(\hat{\pi}_{ppswr}) = \frac{1}{nM_0^2}\sum_{i=1}^{n}p_i\left[\frac{M_i\hat{\pi}_i}{p_i} - M_0\hat{\pi}_{ppswr}\right]^2$$
$$+ \frac{1}{nM_0^2}\sum_{i=1}^{n}\frac{M_i^2}{p_i(m_i-1)}\left[\hat{\pi}_i(1-\hat{\pi}_i) - \frac{\hat{\pi}_i}{\psi_i^2}A_i + \frac{1}{\psi_i^2}B_i\right]. \quad (11)$$

□

If the FSUs are selected proportional to size with M_i, then $p_i = M_i/M_0$. For this reason, we call this method "probability proportional to size" (PPS) sampling. When a sample of the FSU is selected by PPS sampling with replacement via sampling probability, $p_i = M_i/M_0$ for the ith cluster, and m_i SSU are selected by SRSWR from each FSU. The estimator $\hat{\pi}_{ppswr}$ of π is given by:

$$\hat{\pi}_{ppswr} = \frac{1}{n}\sum_{i=1}^{n}\hat{\pi}_i$$
$$= \frac{1}{n}\sum_{i=1}^{n}\frac{1}{m_i}\left[\pi_i(1-\pi_i) + \frac{\pi_i}{\psi_i^2}A_i + \frac{1}{\psi_i^2}B_i\right], \quad (12)$$

and the variance of $\hat{\pi}_{ppswr}$ and its estimator are as follows:

$$V(\hat{\pi}_{ppswr}) = \frac{1}{nM_0}\sum_{i=1}^{N}M_i(\pi_i - \pi)^2$$
$$+ \frac{1}{nM_0}\sum_{i=1}^{N}\frac{M_i}{m_i}\left[\pi_i(1-\pi_i) + \frac{\pi_i}{\psi_i^2}A_i + \frac{1}{\psi_i^2}B_i\right], \quad (13)$$

$$\hat{V}(\hat{\pi}_{ppswr}) = \frac{1}{nM_0} \sum_{i=1}^{n} M_i(\hat{\pi}_i - \hat{\pi}_{ppswr})^2 \qquad (14)$$
$$+ \frac{1}{nM_0} \sum_{i=1}^{n} \frac{M_i}{m_i - 1} \left[\hat{\pi}_i(1 - \hat{\pi}_i) + \frac{\hat{\pi}_i}{\psi_i^2} A_i + \frac{1}{\psi_i^2} B_i \right].$$

2.2. The PPS without Replacement

In this subsection, we consider PPS sampling without replacement to estimate the true population proportion of a sensitive character by applying Yennum et al.'s model, in which n FSUs are drawn by PPS sampling without replacement from the population of N clusters with M_i elementary units for the ith cluster, and m_i SSUs are drawn by SRSWR from each FSU.

From this two-stage sampling, the estimator $\hat{\pi}_{ppswor}$ of π is:

$$\hat{\pi}_{ppswor} = \frac{1}{M_0} \sum_{i=1}^{n} \frac{M_i \hat{\pi}_i}{\theta_i}, \qquad (15)$$

where θ_i is the first inclusion probability for the ith cluster.

The variance of $\hat{\pi}_{ppswor}$ is given by:

$$V(\hat{\pi}_{ppswor}) = \frac{1}{M_0^2} \sum_{i=1}^{N} \sum_{j>i}^{N} (\theta_i \theta_j - \theta_{ij}) \left[\frac{M_i \pi_i}{\theta_i} - \frac{M_j \pi_j}{\theta_j} \right]^2$$
$$+ \frac{1}{M_0^2} \sum_{i=1}^{N} \frac{M_i^2}{m_i \theta_i} \left[\pi_i(1 - \pi_i) + \frac{\pi_i}{\psi_i^2} A_i + \frac{1}{\psi_i^2} B_i \right], \qquad (16)$$

where θ_{ij} is the second inclusion probability of the ith and jth clusters.

Furthermore, the variance estimator of $\hat{\pi}_{ppswor}$ is as follows:

$$\hat{V}(\hat{\pi}_{ppswor}) = \frac{1}{M_0^2} \sum_{i=1}^{n} \sum_{j>i}^{n} \frac{(\theta_i \theta_j - \theta_{ij})}{\theta_{ij}} \left[\frac{M_i \hat{\pi}_i}{\theta_i} - \frac{M_j \hat{\pi}_j}{\theta_j} \right]^2$$
$$+ \frac{1}{M_0^2} \sum_{i=1}^{n} \frac{M_i^2}{\theta_i(m_i - 1)} \left[\hat{\pi}_i(1 - \hat{\pi}_i) + \frac{\hat{\pi}_i}{\psi_i^2} A_i + \frac{1}{\psi_i^2} B_i \right]. \qquad (17)$$

2.3. Two-Stage Equal Probability Sampling

In this subsection, we consider a two-stage equal probability sampling design to estimate the true population proportion of a sensitive characteristic by applying Yennum et al.'s model, in which n FSUs are drawn by simple random sampling without replacement (SRSWOR) from a population of N clusters with M_i elementary units for the ith cluster, and m_i SSUs are drawn by SRSWR from each FSU.

From this two-stage sampling, the estimator $\hat{\pi}_{wr}$ of π is given by:

$$\hat{\pi}_{wr} = \frac{N}{nM_0} \sum_{i=1}^{n} M_i \hat{\pi}_i, \qquad (18)$$

where $\hat{\pi}_i$ is an estimator of the true population proportion for a sensitive characteristic for the ith cluster, which is the same as (4).

The variance of $\hat{\pi}_{wr}$ and its estimator are given as:

$$V(\hat{\pi}_{wr}) = \frac{N^2}{nM_0^2} \frac{1}{(N-1)} \sum_{i=1}^{N} (M_i \pi_i - \overline{M\pi})^2$$
$$+ \frac{N}{nM_0^2} \sum_{i=1}^{N} \frac{M_i^2}{m_i} \left[\pi_i(1 - \pi_i) + \frac{\pi_i}{\psi_i^2} A_i + \frac{1}{\psi_i^2} B_i \right], \qquad (19)$$

$$\hat{V}(\hat{\pi}_{wr}) = \frac{N^2}{nM_0^2} \frac{1}{(n-1)} \sum_{i=1}^{n} (M_i\hat{\pi}_i - \overline{M}\hat{\pi}_{wr})^2$$
$$+ \frac{N}{nM_0^2} \sum_{i=1}^{n} \frac{M_i^2}{m_i - 1} \left[\hat{\pi}_i(1-\hat{\pi}_i) + \frac{\hat{\pi}_i}{\psi_i^2} A_i + \frac{1}{\psi_i^2} B_i \right], \qquad (20)$$

where $\overline{M} = M_0/N$.

3. An Estimation of Sensitive Attributes with Probability Proportional to Size Sampling Under Yennum et al.'s Generalized Model

We consider Yennum et al.'s generalized model, in which generalized geometric distribution is used as a randomization device when n clusters are sampled by PPS sampling or equal probability sampling from the population, which consists of N clusters with size $M_i (i = 1, 2, \cdots, N)$, and $m_i (i = 1, 2, \cdots, n)$ units are drawn by simple random sampling from each sampled cluster.

We develop the sampling schemes for PPS sampling with replacement in Section 3.1 and those for PPS sampling without replacement in Section 3.2. Finally, equal probability sampling is presented in Section 3.3.

3.1. PPS Sampling with Replacement

Let the population be composed of N clusters. In the first stage, a sample of n FSUs is drawn by replacement with the sampling probability p_i for the ith cluster. In the second stage, m_i SSUs are selected by SRSWR from each FSU and guided to apply Yennum et al.'s generalized randomization device.

If the interviewees in the ith cluster choose a white ball during the first stage, and if they have a sensitive attribute A (or A^c), then they are guided to pick replacement balls from the first urn of the second stage device until they take k_{i2} (or k_{i1}) successive balls with their actual status for the first time and are then asked to determine the total number of balls as X_{i1} (or X_{i2}).

If the interviewee in the ith cluster draws a black ball in the first stage, and if they have a sensitive attribute A^c (or A), then they are guided to take replacement balls from the second urn of the second stage device until they take k_{i2} (or k_{i1}) successive balls with their actual status for the first time and are then asked to determine the total number of balls as Y_{i1} (or Y_{i2}).

For the ith cluster, using the randomization device in Figure 1, the total number of balls taken by interviewees X_{i1}, X_{i2}, Y_{i1}, and Y_{i2} are distributed via generalized geometric distribution. Let π_i and $1 - \pi_i$ be the true population proportion of persons who have a sensitive attribute A and A^c for the ith cluster. Assume that each interviewee in the ith cluster is drawn by SRSWR.

For the jth surveyed answer in the ith cluster, Z_{ij} can be expressed as:

$$Z_{ij} = \begin{cases} X_{i1} \text{ with probability} & W_i \pi_i, \\ Y_{i2} \text{ with probability} & (1 - W_i)\pi_i, \\ X_{i2} \text{ with probability} & W_i(1 - \pi_i), \\ Y_{i1} \text{ with probability} & (1 - W_i)(1 - \pi_i), \end{cases} \qquad (21)$$

The expected value of Z_{ij} is given by:

$$E(Z_{ij}) = W_i \pi_i E(X_{i1}) + \pi_i(1 - W_i)E(Y_{i2}) + (1 - \pi_i)W_i E(X_{i2}) + (1 - W_i)(1 - \pi_i)E(Y_{i1})$$
$$= \pi_i \left[W_i \left\{ \frac{1 - P_i^{k_{i1}}}{(1-P_i)P_i^{k_{i1}}} - \frac{1-(1-P_i)^{k_{i2}}}{P_i(1-P_i)^{k_{i2}}} \right\} + (1 - W_i) \left\{ \frac{1-(1-T_i)^{k_{i1}}}{T_i(1-T_i)^{k_{i1}}} - \frac{1 - T_i^{k_{i2}}}{(1-T_i)T_i^{k_{i2}}} \right\} \right]$$
$$+ W_i \left\{ \frac{1-(1-P_i)^{k_{i2}}}{P_i(1-P_i)^{k_{i2}}} \right\} + (1 - W_i) \left\{ \frac{1 - T_i^{k_{i2}}}{(1-T_i)T_i^{k_{i2}}} \right\}. \qquad (22)$$

Then, the formula (22) can be expressed as:

$$E(Z_{ij}) = W_i\left\{\frac{1-(1-P_i)^{k_{i2}}}{P_i(1-P_i)^{k_{i2}}}\right\} - (1-W_i)\left\{\frac{1-T_i^{k_{i2}}}{(1-T_i)T_i^{k_{i2}}}\right\}$$

$$= T_i\left[W_i\left\{\frac{1-P_i^{k_{i1}}}{(1-P_i)P_i^{k_{i1}}} - \frac{1-(1-P_i)^{k_{i2}}}{P_i(1-P_i)^{k_{i2}}}\right\} + (1-W_i)\left\{\frac{1-(1-T_i)^{k_{i2}}}{T_i(1-T_i)^{k_{i2}}} - \frac{1-T_i^{k_{i1}}}{(1-T_i)T_i^{k_{i1}}}\right\}\right]. \qquad (23)$$

The estimator $\hat{\pi}_{iG}$ of the population proportion π_i for the ith cluster is given by:

$$\hat{\pi}_{iG} = \frac{(1-T_i)^{k_{i1}+1}T_i^{k_{i2}+1}(1-P_i)^{k_{i2}+1}P_i^{k_{i1}+1}}{m_i\varphi_{i2}}\left(\sum_{j=1}^{m_i} Z_{ij} - \varphi_{i1}\right), \qquad (24)$$

where:

$$\varphi_{i1} = W_i\{1-(1-P_i)^{k_{i2}}\}(1-T_i)^{k_{i1}+1}T_i^{k_{i2}+1}(1-P_i)P_i^{k_{i1}} \\ + (1-W_i)(1-T_i^{k_{i2}})P_i^{k_{i1}+1}(1-P_i)^{k_{i2}+1}T_i(1-T_i)^{k_{i1}+1}, \qquad (25)$$

and:

$$\varphi_{i2} = W_i\left[(1-P_i)^{k_{i1}}P_iT_i^{k_{i2}+1}(1-T_i)^{k_{i1}+1}(1-P_i)^{k_{i2}} - \{1-(1-P_i)^{k_{i2}}\}P_i^{k_{i1}}(1-P_i)T_i^{k_{i2}+1}(1-T_i)^{k_{i1}+1}\right] \\ + (1-W_i)\left[\{1-(1-T_i)^{k_{i1}}\}P_i^{k_{i1}+1}(1-T_i)(1-P_i)^{k_{i2}+1}T_i^{k_{i2}} - (1-T_i^{k_{i2}})P_i^{k_{i1}+1}T_i(1-P_i)^{k_{i2}+1}(1-T_i)^{k_{i1}}\right]. \qquad (26)$$

When the interviewees are sampled by SRSWR for the ith cluster selected with a replacement by sampling probability p_i, the estimator $\hat{\pi}_{Gppswr}$ of the true population proportion π of a sensitive attribute is:

$$\hat{\pi}_{Gppswr} = \frac{1}{nM_0}\sum_{i=1}^{n}\frac{M_i\hat{\pi}_{iG}}{p_i} \\ = \frac{1}{nM_0}\sum_{i=1}^{n}\frac{M_i}{p_i}\left[\frac{(1-T_i)^{k_{i1}+1}T_i^{k_{i2}+1}(1-P_i)^{k_{i2}+1}P_i^{k_{i1}+1}}{m_i\varphi_{i2}}\left(\sum_{j=1}^{m_i}Z_{ij} - \varphi_{i1}\right)\right], \qquad (27)$$

where $M_0 = \sum_{i=1}^{N} M_i$.

Theorem 3: The estimator $\hat{\pi}_{Gppswr}$ of the true population proportion π of a sensitive character is an unbiased estimator.

Proof:

$$E_1E_2(\hat{\pi}_{Gppswr}) = E_1E_2\left[\frac{1}{nM_0}\sum_{i=1}^{n}\frac{M_i\hat{\pi}_{iG}}{p_i}\right] = E_1\left[\frac{1}{nM_0}\sum_{i=1}^{n}\frac{M_iE_2(\hat{\pi}_{iG})}{p_i}\right],$$

and, since:

$$E_2(\hat{\pi}_{iG}) = E_2\left[\frac{(1-T_i)^{k_{i1}+1}T_i^{k_{i2}+1}(1-P_i)^{k_{i2}+1}P_i^{k_{i1}+1}}{m_i\varphi_{i2}}\left(\sum_{j=1}^{m_i}Z_{ij} - \varphi_{i1}\right)\right]$$

$$= \frac{(1-T_i)^{k_{i1}+1}T_i^{k_{i2}+1}(1-P_i)^{k_{i2}+1}P_i^{k_{i1}+1}}{m_i\varphi_{i2}}\left(\sum_{j=1}^{m_i}E_2(Z_{ij}) - \varphi_{i1}\right)$$

$$= \pi_i,$$

we can obtain:

$$E_1E_2(\hat{\pi}_{Gppswr}) = E_1\left[\frac{1}{nM_0}\sum_{i=1}^{n}\frac{M_i\pi_i}{p_i}\right] = \frac{1}{M_0}\sum_{i=1}^{N}p_i\frac{M_i\pi_i}{p_i} = \pi.$$

□

Theorem 4: The variance of $\hat{t}_{Gppswr}$ is obtained by a two-stage sampling scheme, such that a sample of n FSU is selected with replacement by sampling probability p_i for the ith cluster from the population of N clusters consisting of M_i elements for the ith cluster, and m_i SSUs are drawn by SRSWR from each FSU, as given by:

$$V(\hat{t}_{Gppswr}) = \frac{1}{nM_0^2} \sum_{i=1}^{N} p_i \left[\frac{M_i \pi_i}{p_i} - M_0 \pi \right]^2 \\ + \frac{1}{nM_0^2} \sum_{i=1}^{N} \frac{M_i^2}{m_i p_i} \left[\frac{\left\{(1-T_i)^{k_{i1}+1} T_i^{k_{i2}+1}(1-P)^{k_{i2}+1} P_i^{k_{i1}+1}\right\}^2}{\varphi_{i2}^2} \sigma_{iZ}^2 \right], \tag{28}$$

where:

$$\sigma_{iZ}^2 = E(Z_{ij}^2) - (E(Z_{ij}))^2$$

$$= \pi_i \left[W_i \left(\frac{1-(2k_{i1}+1)(1-P_i)P_i^{k_{i1}} - P_i^{2k_{i1}+1} + (1-P_i^{k_{i1}})^2}{(1-P_i)^2 P_i^{2k_{i1}}} \right) \right.$$
$$\left. + (1-W_i) \left(\frac{1-(2k_{i1}+1)T_i(1-T_i)^{k_{i1}} - (1-T_i)^{2k_{i1}+1} + (1-(1-T_i)^{k_{i1}})^2}{T_i^2(1-T_i)^{2k_{i1}}} \right) \right]$$
$$+ (1-\pi_i) \left[W_i \left(\frac{1-(2k_{i2}+1)P_i(1-P_i)^{k_{i2}} - (1-P_i)^{2k_{i2}+1} + (1-(1-P_i^{k_{i2}}))^2}{P_i^2(1-P_i)^{2k_{i2}}} \right) \right.$$
$$\left. + (1-W_i) \left(\frac{1-(2k_{i2}+1)(1-T_i)T_i^{k_{i2}} - T_i^{2k_{i2}+1} + (1-T_i^{k_{i2}})^2}{(1-T_i)^2 T_i^{2k_{i2}}} \right) \right]$$
$$- \left[\pi_i \left\{ W_i \left(\frac{1-P_i^{k_{i1}}}{(1-P_i)P_i^{k_{i1}}} - \frac{1-(1-P_i)^{k_{i2}}}{P_i(1-P_i)^{k_{i2}}} \right) + (1-W_i) \left(\frac{1-(1-T_i)^{k_{i1}}}{T_i(1-T_i)^{k_{i1}}} - \frac{1-T_i^{k_{i2}}}{(1-T_i)T_i^{k_{i2}}} \right) \right\} \right.$$
$$\left. + (1-\pi_i) \left\{ W_i \left(\frac{1-(1-P_i)^{k_{i2}}}{P_i(1-P_i)^{k_{i2}}} \right) + (1-W_i) \left(\frac{1-T_i^{k_{i2}}}{(1-T_i)T_i^{k_{i2}}} \right) \right\} \right]^2. \tag{29}$$

Proof: The total number of balls taken by interviewees for the ith cluster, X_{i1}, X_{i2}, Y_{i1} and Y_{i2}, are random variables with variances:

$$V(X_{i1}) = \frac{1-(2k_{i1}+1)(1-P_i)P_i^{k_{i1}} - P_i^{2k_{i1}+1}}{(1-P_i)^2 P_i^{2k_{i1}}}, \tag{30}$$

$$V(X_{i2}) = \frac{1-(2k_{i2}+1)P_i(1-P_i)^{2k_{i2}} - (1-P_i)^{2k_{i2}+1}}{P_i^2(1-P_i)^{2k_{i2}}}, \tag{31}$$

$$V(Y_{i1}) = \frac{1-(2k_{i2}+1)(1-T_i)T_i^{k_{i2}} - T_i^{2k_{i2}+1}}{(1-T_i)^2 T_i^{2k_{i2}}}, \tag{32}$$

$$V(Y_{i2}) = \frac{1-(2k_{i1}+1)T_i(1-T_i)^{k_{i1}} - (1-T_i)^{2k_{i1}+1}}{T_i^2(1-T_i)^{2k_{i1}}}. \tag{33}$$

From (21), to drive the variance of $\hat{t}_{Gppswr}$ we can obtain the expected values of Z_{ij} and Z_{ij}^2 as follows:

$$E(Z_{ij}) = \pi_i \left[W_i \left(\frac{1-P_i^{k_{i1}}}{(1-P_i)P_i^{k_{i1}}} \right) + (1-W_i) \left(\frac{1-(1-T_i)^{k_{i1}}}{T_i(1-T_i)^{k_{i1}}} \right) \right] \\ + (1-\pi_i) \left[W_i \left(\frac{1-(1-P_i)^{k_{i2}}}{P_i(1-P_i)^{k_{i2}}} \right) + (1-W_i) \left(\frac{1-T_i^{k_{i2}}}{(1-T_i)T_i^{k_{i2}}} \right) \right], \tag{34}$$

$$
\begin{aligned}
E(Z_{ij}^2) &= \pi_i\left[W_i E(X_{i1}^2) + (1-W_i)E(Y_{i1}^2)\right] + (1-\pi_i)\left[W_i E(X_{i2}^2) + (1-W_i)E(Y_{i2}^2)\right] \\
&= \pi_i\left[W_i\left(\frac{1-(2k_{i1}+1)(1-P_i)P_i^{k_{i1}} - P_i^{2k_{i1}+1} + (1-P_i^{k_{i1}})^2}{(1-P_i)^2 P_i^{2k_{i1}}}\right)\right. \\
&\quad + (1-W_i)\left(\frac{1-(2k_{i1}+1)T_i(1-T_i)^{k_{i1}} - (1-T_i)^{2k_{i1}+1} + \{1-(1-T_i)^{k_{i1}}\}^2}{T_i^2(1-T_i)^{2k_{i1}}}\right) \Bigg] \\
&\quad + (1-\pi_i)\left[W_i\left(\frac{1-(2k_{i2}+1)P_i(1-P_i)^{k_{i2}} - (1-P_i)^{2k_{i2}+1} + \{1-(1-P_i^{k_{i2}})\}^2}{P_i^2(1-P_i)^{2k_{i2}}}\right)\right. \\
&\quad + (1-W_i)\left(\frac{1-(2k_{i2}+1)(1-T_i)T_i^{k_{i2}} - T_i^{2k_{i2}+1} + (1-T_i^{k_{i2}})^2}{(1-T_i)^2 T_i^{2k_{i2}}}\right) \Bigg].
\end{aligned}
\tag{35}
$$

Since $V(\hat{\pi}_{Gppswr}) = V_1 E_2(\hat{\pi}_{Gppswr}) + E_1 V_2(\hat{\pi}_{Gppswr})$,

$$
\begin{aligned}
V_1 E_2(\hat{\pi}_{Gppswr}) &= V_1 E_2\left[\frac{1}{nM_0}\sum_{i=1}^{n}\frac{M_i \hat{\pi}_{iG}}{p_i}\right] \\
&= V_1\left[\frac{1}{nM_0}\sum_{i=1}^{n}\frac{M_i \pi_i}{p_i}\right] \\
&= \frac{1}{nM_0^2}\sum_{i=1}^{N} p_i\left[\frac{M_i \pi_{iG}}{p_i} - M_0\pi\right]^2,
\end{aligned}
$$

and:

$$
\begin{aligned}
E_1 V_2(\hat{\pi}_{Gppswr}) &= E_1 V_2\left[\frac{1}{nM_0}\sum_{i=1}^{n}\frac{M_i \hat{\pi}_{iG}}{p_i}\right] \\
&= E_1\left[\frac{1}{(nM_0)^2}\sum_{i=1}^{n}\frac{M_i^2}{p_i^2} V_2(\hat{\pi}_{iG})\right] \\
&= E_1\left[\frac{1}{(nM_0)^2}\sum_{i=1}^{n}\frac{M_i^2}{p_i^2} V_2\left\{\frac{(1-T_i)^{k_{i1}+1}T_i^{k_{i2}+1}(1-P_i)^{k_{i2}+1}P_i^{k_{i1}+1}}{m_i \varphi_{i2}}\left(\sum_{j=1}^{m_i}Z_{ij} - \varphi_{i1}\right)\right\}\right] \\
&= E_1\left[\frac{1}{(nM_0)^2}\sum_{i=1}^{n}\frac{M_i^2}{p_i^2}\frac{1}{m_i}\left\{\frac{\{(1-T_i)^{k_{i1}+1}T_i^{k_{i2}+1}(1-P_i)^{k_{i2}+1}P_i^{k_{i1}+1}\}^2}{\varphi_{i2}^2}\sigma_{iZ}^2\right\}\right] \\
&= \frac{1}{nM_0^2}\sum_{i=1}^{N}\frac{M_i^2}{m_i p_i}\left[\frac{\{(1-T_i)^{k_{i1}+1}T_i^{k_{i2}+1}(1-P_i)^{k_{i2}+1}P_i^{k_{i1}+1}\}^2}{\varphi_{i2}^2}\sigma_{iZ}^2\right].
\end{aligned}
$$

We can then obtain the variance (28). Also, an unbiased estimator of $V(\hat{\pi}_{Gppswr})$ is given by:

$$
\begin{aligned}
\hat{V}(\hat{\pi}_{Gppswr}) &= \frac{1}{nM_0^2}\sum_{i=1}^{n} p_i\left[\frac{M_i \hat{\pi}_{iG}}{p_i} - M_0\hat{\pi}_{Gppswr}\right]^2 \\
&\quad + \frac{1}{nM_0^2}\sum_{i=1}^{n}\frac{M_i^2}{p_i(m_i-1)}\left[\frac{\{(1-T_i)^{k_{i1}+1}T_i^{k_{i2}+1}(1-P)^{k_{i2}+1}P_i^{k_{i1}+1}\}^2}{\varphi_{i2}^2}\hat{\sigma}_{iZ}^2\right].
\end{aligned}
\tag{36}
$$

□

3.2. PPS Sampling Without Replacement

In this subsection, we consider PPS sampling without replacement to estimate the true population proportion of a sensitive characteristic by applying Yennum et al.'s generalized model, in which n FSUs are drawn by PPS sampling without replacement from a population of N clusters with M_i elementary units for the ith cluster, and m_i SSUs are drawn by SRSWR from each FSU.

From this procedure, the estimator $\hat{\pi}_{Gppswor}$ of π is given by:

$$\hat{\pi}_{Gppswor} = \frac{1}{M_0}\sum_{i=1}^{n}\frac{M_i\hat{\pi}_{iG}}{\theta_i}, \tag{37}$$

where θ_i is the first inclusion probability for the ith cluster.

The variance of $\hat{\pi}_{Gppswor}$ is given by:

$$V(\hat{\pi}_{Gppswor}) = \frac{1}{M_0^2}\sum_{i=1}^{N}\sum_{j>i}^{N}(\theta_i\theta_j - \theta_{ij})\left[\frac{M_i\pi_i}{\theta_i} - \frac{M_j\pi_j}{\theta_j}\right]^2$$
$$+ \frac{1}{M_0^2}\sum_{i=1}^{N}\frac{M_i^2}{m_i\theta_i}\left[\frac{\{(1-T_i)^{k_{i1}+1}T_i^{k_{i2}+1}(1-P)^{k_{i2}+1}P_i^{k_{i1}+1}\}^2}{\varphi_{i2}^2}-\sigma_{iZ}^2\right], \tag{38}$$

where θ_{ij} is the second inclusion probability for ith and jth clusters.

Also, the variance estimator of $\hat{\pi}_{Gppswor}$ is:

$$\hat{V}(\hat{\pi}_{Gppswor}) = \frac{1}{M_0^2}\sum_{i=1}^{n}\sum_{j>i}^{n}\frac{(\theta_i\theta_j-\theta_{ij})}{\theta_{ij}}\left[\frac{M_i\hat{\pi}_{iG}}{\theta_i} - \frac{M_j\hat{\pi}_{jG}}{\theta_j}\right]^2$$
$$+\frac{1}{M_0^2}\sum_{i=1}^{n}\frac{M_i^2}{\theta_i(m_i-1)}\left[\frac{\{(1-T_i)^{k_{i1}+1}T_i^{k_{i2}+1}(1-P)^{k_{i2}+1}P_i^{k_{i1}+1}\}^2}{\varphi_{i2}^2}-\hat{\sigma}_{iZ}^2\right]. \tag{39}$$

3.3. Two-Stage Equal Probability Sampling

In this subsection, we consider a two-stage equal probability sampling scheme to estimate the true population proportion of a sensitive attribute by applying Yennum et al.'s generalized model, in which n FSUs are drawn by SRSWOR from a population of N clusters consisting of M_i elementary units for the ith cluster, and m_i SSUs are drawn by SRSWR from each FSU.

From this procedure, the estimator $\hat{\pi}_{Gwr}$ of the true population proportion π for a sensitive attribute is given by:

$$\hat{\pi}_{Gwr} = \frac{N}{nM_0}\sum_{i=1}^{n}M_i\hat{\pi}_{iG}, \tag{40}$$

where the estimator $\hat{\pi}_{iG}$ is the estimator of a sensitive characteristic of the ith cluster, which is the same as (24).

The variance and variance estimator of $\hat{\pi}_{Gwr}$ are:

$$V(\hat{\pi}_{Gwr}) = \frac{N^2}{nM_0^2}\sum_{i=1}^{N}\frac{1}{N-1}\left[M_i\pi_i - \overline{M\pi}\right]^2$$
$$+\frac{N}{nM_0^2}\sum_{i=1}^{N}\frac{M_i^2}{m_i}\left[\frac{\{(1-T_i)^{k_{i1}+1}T_i^{k_{i2}+1}(1-P)^{k_{i2}+1}P_i^{k_{i1}+1}\}^2}{\varphi_{i2}^2}-\sigma_{iZ}^2\right], \tag{41}$$

and:

$$\hat{V}(\hat{\pi}_{Gwr}) = \frac{N^2}{nM_0^2}\sum_{i=1}^{n}\frac{1}{n-1}(M_i\hat{\pi}_{iG} - \overline{M}\hat{\pi}_{Gwr})^2$$
$$+\frac{N}{nM_0^2}\sum_{i=1}^{N}\frac{M_i^2}{m_i-1}\left[\frac{\{(1-T_i)^{k_{i1}+1}T_i^{k_{i2}+1}(1-P)^{k_{i2}+1}P_i^{k_{i1}+1}\}^2}{\varphi_{i2}^2}-\hat{\sigma}_{iZ}^2\right], \tag{42}$$

respectively, where $\overline{M} = M_0/N$.

4. Efficiency Comparisons

4.1. PPSWR Sampling versus Equal Probability Two-Stage Sampling in Yennum et al.'s Model

If we assume $N - 1 \doteq N$, then the difference between the variance of equal probability two-stage sampling, (19), and the variance of PPS with replacement sampling, (6), is given by:

$$V(\hat{\pi}_{wr}) - V(\hat{\pi}_{ppswr}) = \frac{1}{nM_0\overline{M}} \left[\sum_{i=1}^{N} (M_i - \overline{M})^2 \pi_i^2 + \overline{M} \left\{ \sum_{i=1}^{N} (M_i - \overline{M})(\pi_i^2 - \pi^2) \right\} \right. $$
$$+ \sum_{i=1}^{N} (M_i - \overline{M})^2 \frac{1}{m_i} \left(\pi_i(1 - \pi_i) + \frac{\pi_i}{\psi_i^2} A_i + \frac{1}{\psi_i^2} B \right) \qquad (43)$$
$$\left. + \overline{M} \left\{ \sum_{i=1}^{N} (M_i - \overline{M}) \frac{1}{m_i} \left(\pi_i(1 - \pi_i) + \frac{\pi_i}{\psi_i^2} A_i + \frac{1}{\psi_i^2} B \right) \right\} \right].$$

In (43), we can see that $V(\hat{\pi}_{wr}) = V(\hat{\pi}_{ppswr})$ under the condition $M_i = \overline{M} = M_0/N$; i.e., if the cluster sizes are equal, the selection probabilities of the PPS with replacement sampling are all N^{-1} and equal to those of equal probability two-stage replacement sampling.

If the size of a cluster, M_i is significantly different, then $\sum_{i=1}^{N} (M_i - \overline{M})^2 \pi_i^2$, the first term on the right side of (43), has large values, and the second term, $\sum_{i=1}^{N} (M_i - \overline{M})^2 (\pi_i^2 - \pi^2)$, has relatively small values. Hence, the estimation by PPS with replacement sampling is more efficient than that by equal probability two-stage replacement sampling.

We used the relative efficiency (RE) to compare the efficiency of the two sampling methods—PPS with replacement sampling and equal probability two-stage replacement sampling:

$$RE_1 = \frac{V(\hat{\pi}_{wr})}{V(\hat{\pi}_{ppswr})} \times 100(\%).$$

Values of RE_1 over 100% indicate that the estimator obtained by the PPS with the replacement sampling method was more efficient than the estimator obtained by the equal probability two-stage replacement sampling.

In calculating REs, we set the parameters as follows:

$M_0 = 10,000, M_1 = 1,000, M_2 = 2,000, M_3 = 3,000, M_4 = 4,000$
$m_0 = 1,000, m_1 = 100, m_2 = 200, m_3 = 300, m_4 = 400,$
$p_1 = 0.235, p_2 = 0.441, p_3 = 0.609, p_4 = 0.715.$

From Table 1, when the selection probability W for the first-stage randomization device increased from 0.1 to 0.9 by 0.2 and the second stage randomization devices T increased from 0.6 to 0.8 by 0.1 and P from 0.65 to 0.90 by 0.05, REs increase under the fixed proportion of a sensitive attribute (particularly when the selection probability of the second randomization device T increased), and the RE increased according to the conditions of P and π_i.

Table 1. The relative efficiencies (REs) of a sensitive estimator between the probability proportional to size (PPS) sampling with replacement and the equal probability two-stage sampling with replacement in Yennum et al.'s model to change π_i and W.

π_i	T	W=0.1			W=0.3			W=0.5			W=0.7			W=0.9		
	P	0.6	0.7	0.8	0.6	0.7	0.8	0.6	0.7	0.8	0.6	0.7	0.8	0.6	0.7	0.8
0.1	0.65	56.59	95.07	123.5	48.08	61.06	91.71	52.73	46.18	55.97	63.18	52.59	39.63	75.51	71.08	61.88
	0.7	54.42	89.17	120	50	54.61	81.69	58.65	48.28	48.17	71.01	60.85	45.08	83.88	80.2	72.61
	0.75	52.93	81.67	114.8	53.61	51.26	70.48	64.76	53.56	45.34	77.67	69.08	54.13	90.33	87.46	81.68
	0.8	52.61	72.72	106.4	58.15	51.63	59.9	70.32	60.37	48.28	83.01	76.38	64.29	95.2	93.08	88.97
	0.85	53.84	63.67	93.17	62.84	55.3	53.48	75.02	67.37	55.79	87.17	82.48	73.91	98.87	97.41	94.69
	0.9	56.57	57.65	74.27	67.15	61.05	54.47	78.79	73.8	65.52	90.32	87.41	82.23	101.6	100.7	99.15
0.2	0.65	82.74	134.4	153.1	50.22	92.87	130.8	60.29	48.44	86.05	90.64	61.69	34.18	117.4	108.9	89.02
	0.7	75.82	129.7	151.4	51.78	76.68	121.8	77.73	48.28	65.5	108.7	84.97	43.35	130.7	125.1	111.8
	0.75	68.56	122.7	148.7	60.97	61.93	108.4	94.15	62.33	48.62	121.2	104.6	67.95	139	135.4	127
	0.8	62.59	111.6	144.1	74.34	56.05	88.85	107.2	82.13	48.28	129.7	118.9	94	144.4	142.1	137
	0.85	61.04	94.57	135	87.93	64.32	66.55	116.7	100.4	68.61	135.4	128.9	114.4	148	146.5	143.6
	0.9	66.6	74.23	114.3	99.4	82.02	61.52	123.5	114.4	95.72	139.3	135.8	128.5	150.4	149.6	148.1
0.3	0.65	106.8	152.4	164.9	54.34	117.7	149.2	70.41	53.88	109.9	119	74.13	31.19	148.1	139.5	115.9
	0.7	98.24	149.4	163.9	53.64	99.37	142.7	100.6	48.28	85.26	139.9	111.1	42.92	159	154.4	141.5
	0.75	87.63	144.7	162.4	70.27	76.57	132.3	124	73.92	55.9	151.7	135.4	85.58	164.8	162.2	155.2
	0.8	76.17	136.6	159.9	94.79	61.53	114	139.2	107.4	48.28	158.6	149.6	122.7	168.2	166.6	162.8
	0.85	69.9	121.6	154.7	116.5	75.47	83.46	148.7	131.9	85.14	162.8	158	145.2	170.2	169.4	167.4
	0.9	78.83	94.88	140.8	131.9	107.7	69.61	154.7	146.9	126.4	165.4	163.1	157.6	171.6	171.1	170.2
0.4	0.65	124.6	162.1	171.2	59.97	134.4	159.3	82.29	61.29	126.3	141.7	88.14	30.06	166.3	159.3	136.7
	0.7	116.5	160	170.5	55.58	117.6	154.6	122.6	48.28	102.2	160.3	133.3	43.6	173.7	170.4	159.7
	0.75	105.4	156.9	169.6	81.25	91.92	146.7	147.4	87.35	65.18	169.2	156.3	103.3	177.3	175.5	170.3
	0.8	91.07	151.5	168	116.6	67.99	131.8	160.8	130.8	48.28	173.8	167.6	144.2	179.1	178.2	175.5
	0.85	80.35	140.6	164.9	142.3	88.79	100.6	168.2	155.2	103.2	176.4	173.5	164	180.2	179.7	178.4
	0.9	93.87	116	156.4	157	133.7	78.99	172.3	167.2	150.6	178	176.6	173.2	180.9	180.6	180.1

On the other hand, RE increased when the first-stage selection probability W was less than 0.5, and the values of T, P, and π_i (from 0.1 to 0.4) decreased, but the RE decreased when the value of W was greater than 0.5 under a fixed value for T, P, and π_i.

Furthermore, the greater the true population proportion of a sensitive attribute π_i, the higher the overall efficiency of Yennum et al.'s model, as shown by the values of the bottom cells in Table 1. This result agrees with the typical sampling survey methodology as the true population proportion of a sensitive attribute π_i increases.

4.2. PPSWR Sampling versus Equal Probability Two-Stage Sampling in Yennum et al.'s Generalized Model

If we assume $N - 1 \doteq N$, then the difference between the variance of equal probability two-stage sampling scheme (41) and the variance of the PPS with replacement sampling scheme (28) is given by:

$$V(\hat{\pi}_{Gwr}) - V(\hat{\pi}_{Gppswr}) = \frac{1}{nM_0\overline{M}} \left[\sum_{i=1}^{N} (M_i - \overline{M})^2 \pi_i^2 + \overline{M} \left\{ \sum_{i=1}^{N} (M_i - \overline{M})(\pi_i^2 - \pi^2) \right\} \right.$$
$$+ \sum_{i=1}^{N} (M_i - \overline{M})^2 \frac{1}{m_i} \left[\frac{\left\{(1-T_i)^{k_{i1}+1} T_i^{k_{i2}+1} (1-P)^{k_{i2}+1} P_i^{k_{i1}+1}\right\}^2}{\varphi_{i2}^2} \sigma_{iZ}^2 \right] \quad (44)$$
$$\left. + \overline{M} \left\{ \sum_{i=1}^{N} (M_i - \overline{M}) \frac{1}{m_i} \left(\frac{\left\{(1-T_i)^{k_{i1}+1} T_i^{k_{i2}+1} (1-P)^{k_{i2}+1} P_i^{k_{i1}+1}\right\}^2}{\varphi_{i2}^2} \sigma_{iZ}^2 \right) \right\} \right].$$

In (44), we can see that $V(\hat{\pi}_{Gwr}) = V(\hat{\pi}_{Gppswr})$ under the condition $M_i = \overline{M} = M_0/N$, i.e., if the cluster sizes are equal, the selection probabilities of the PPS with replacement sampling are all N^{-1} and equal to those of the equal probability two-stage replacement sampling.

If cluster sizes, M_i, were significantly different, then $\sum_{i=1}^{N}(M_i - \overline{M})^2 \pi_i^2$, the first term of the right-hand side in (44), had large values, and the second term, $\sum_{i=1}^{N}(M_i - \overline{M})^2(\pi_i^2 - \pi^2)$, had relatively small values. Hence, the estimation by PPS with replacement sampling is more efficient than that by equal probability two-stage replacement sampling.

We used the relative efficiency (RE) to compare the efficiency of the two sampling designs (PPS with replacement sampling and equal probability two-stage replacement sampling):

$$RE_2 = \frac{V(\hat{\pi}_{Gwr})}{V(\hat{\pi}_{Gppswr})} \times 100(\%)$$

Values of RE_2 over 100% indicate that the estimator obtained by PPS with the replacement sampling method was more efficient than the estimator obtained by equal probability two-stage replacement sampling.

Table 2 shows the results of the REs obtained by increasing the true population proportion π_i from 0.1 to 0.4 by 0.1. The selection probabilities of the randomized response model (W, T and P) are shown in Section 4.1.

Table 2. The REs for a sensitive estimator between the PPS with replacement sampling and equal probability two-stage sampling with replacement in Yennum et al.'s generalized model for changing π_i and W.

π_i	T	W 0.1			0.3			0.5			0.7			0.9		
	P	0.6	0.7	0.8	0.6	0.7	0.8	0.6	0.7	0.8	0.6	0.7	0.8	0.6	0.7	0.8
0.1	0.65	171.5	172.8	172.8	163	166.9	167.9	145.4	155.4	159.1	104.1	126.5	138.8	48.17	52.15	68.27
	0.7	167.5	169.7	170.1	152.7	160.2	163.1	119.1	139.1	148.9	57.85	83.09	111	78.98	76.63	58.68
	0.75	162.2	165.9	167	135.6	150.2	156.7	79	111	133.1	54.18	48.22	69.55	116.3	124.4	117.6
	0.8	154.2	160.9	163.3	106.2	133.1	147.2	48.74	68.09	105.6	88.1	76.9	50.14	136.5	148.1	152.8
	0.85	139.8	153.1	158.5	64.63	99.62	130.1	64.15	50.65	60.85	116.4	120.1	102.3	147	159.4	168.3
	0.9	108.4	136	150.3	49.94	51.5	89.01	96.84	95.38	69.51	133	144.7	150.1	152.7	165.1	175.3
0.2	0.65	180.9	181.1	181	177	178.1	178.3	168.4	172	173.1	140.7	154.2	160	48.39	59.25	87.17
	0.7	178.9	179.5	179.5	172.1	174.6	175.5	152.4	162.4	166.8	75.03	112.8	137.8	120.5	111	71.17
	0.75	176.4	177.5	177.8	163.1	169.2	171.8	113.3	142.1	156.1	67.28	48.47	88.43	159.6	161.5	149
	0.8	172.6	174.9	175.8	143.4	159.1	166.1	50.3	91.74	133.4	132.9	110.7	52.83	171.6	175.4	173.9
	0.85	165.8	171	173.1	91.83	133.9	154.7	94.45	55.74	74.87	160	159	135.4	176.4	180.4	182.1
	0.9	147.3	162.3	168.5	55.09	58.5	118.4	145.2	139.7	94.09	170.3	174.6	173.7	178.6	182.6	185.2
0.3	0.65	184	184	183.9	181.6	182	182	176.2	178	178.5	157.3	165.5	168.9	48.68	66.76	101.4
	0.7	182.8	183	182.9	178.5	179.7	180.2	165.6	171.4	174	91.45	130.8	151.3	145.6	132.9	81.96
	0.75	181.2	181.7	181.8	172.9	176.1	177.6	134.5	156.5	166.1	81.89	48.76	102.6	174.4	173.9	161.9
	0.8	178.9	180	180.4	159.7	169.3	173.6	52.38	109.5	147.9	155.9	132.1	55.54	181.2	182.4	180.1
	0.85	174.8	177.5	178.6	114.3	150.9	165.5	120.4	61.6	86.84	174.9	172.4	151.2	183.6	185.3	185.4
	0.9	163.6	172.1	175.6	62.66	66.66	135.6	166.4	160.1	112.1	180.7	182.3	180.3	184.7	186.4	187.3
0.4	0.65	185.5	185.5	185.4	183.9	184.1	184	180	181	181.3	166.1	171.6	173.8	49.04	74.03	112.4
	0.7	184.7	184.8	184.7	181.7	182.4	182.6	172.3	176.1	177.9	105.3	142.5	159.4	159.6	146.4	91.15
	0.75	183.5	183.8	183.9	177.7	179.7	180.7	147.7	164.5	171.7	95.48	49.07	113.7	180.5	179.2	168.8
	0.8	181.9	182.6	182.9	168.2	174.7	177.7	54.79	122.6	156.8	167.6	145.4	58.24	184.8	185.2	182.8
	0.85	179.2	180.8	181.6	130.6	160.5	171.4	138.7	67.61	97.01	181	178.3	160.1	186.3	187	186.7
	0.9	171.7	177.1	179.3	71.76	75	146.8	176	170.1	125	184.7	185.3	183.2	186.9	187.8	188.1

In calculating the REs, we set the parameters as follows:

$$M_0 = 10,000, M_1 = 1,000, M_2 = 2,000, M_3 = 3,000, M_4 = 4,000$$
$$m_0 = 1,000, m_1 = 100, m_2 = 200, m_3 = 300, m_4 = 400,$$
$$p_1 = 0.235, p_2 = 0.441, p_3 = 0.609, p_4 = 0.715,$$
$$k_1 = 2, k_2 = 1.$$

From the results of Table 2, the efficiencies vary according to changes in the probabilities of selection during the first stage W and the second stage T and P in the randomization device, but when the first-stage selection probability W is fixed, and the second-stage selection probabilities T and P increase, then the relative efficiency of the PPS sampling is better than that of the equal probability two-stage sampling in Yennum et al.'s model.

5. Conclusions

We extended Yennum et al.'s model, in which geometric distribution is used as a randomization device for a population consisting of different-sized clusters, and clusters are selected by PPS sampling. Estimators for the true population proportion of a sensitive attribute, their variances, and their variance estimators are derived under PPS sampling and equal probability two-stage sampling.

We also applied these sampling designs to the case of Yennum et al.'s generalized model. Numerical studies were carried out to compare the efficiencies of the proposed methods in each case of Yennum et al.'s model and Yennum et al.'s generalized model in cases with a replacement.

Although the experiments were assumed to use a replacement, we expected similar results for a case without replacement, as per typical sampling theory.

From the numerical study, we found that the efficiency of the two-stage sampling for probability proportional to size depends on the given parameter values, but the efficiency of Yennum et al.'s generalized model is preferred for most combinations of parameters over around 80%.

Author Contributions: Conceptualization, G.-S.L.; methodology, C.-K.S.; writing—original draft preparation, K.-H.H.; writing—review and editing, C.-K.S.; project administration and funding acquisition, G.-S.L.

Funding: This research was supported by the Basic Science Research Program through the National Research Foundation of Korea (NRF) funded by the Ministry of Education (2018R1D1A3B07044007).

Acknowledgments: We would like to thank the anonymous reviewers for their very careful reading and valuable comments/suggestions.

Conflicts of Interest: The authors declare no conflict of interest.

References

1. Warner, S.L. Randomized response: A survey technique for eliminating evasive answer bias. *J. Am. Stat. Assoc.* **1965**, *60*, 63–69. [CrossRef] [PubMed]
2. Cochran, W.G. *Sampling Techniques*, 3rd ed.; John Wiley and Sons: New York, NY, USA, 1977.
3. Fox, J.A.; Tracy, P.E. *Randomized Response: A Method for Sensitive Survey*; Sage Publications: Newbury Park, CA, USA, 1986.
4. Kuk, A.Y.C. Asking sensitive questions indirectly. *Biometrika* **1990**, *77*, 436–438. [CrossRef]
5. Chaudhuri, A.; Mukerjee, R. *Randomized Response: Theory and Techniques*; Marcel Dekker Inc.: New York, NY, USA, 1988.
6. Ryu, J.B.; Hong, K.H.; Lee, G.S. *Randomized Response Model*; Freedom Academy: Seoul, Korean, 1993.
7. Lee, G.S.; Hong, K.H. Randomized response model by two-stage cluster sampling. *Korean Commun. Stat.* **1998**, *5*, 99–105.
8. Lee, G.S. A Study on the Randomized Response Technique by PPS Sampling. *Korean J. Appl. Stat.* **2006**, *19*, 69–80.

9. Yennum, N.Y.; Sedory, S.A.; Singh, S. Improved strategy to collect sensitive data by using geometric distribution as a randomization device. *Commun. Stat. Theory Methods* **2019**, *48*, 5777–5795. [CrossRef]
10. Hussain, Z.; Shabbir, J.; Pervez, Z.; Shah, S.F.; Khan, M. Generalized geometric distribution of order k: A flexible choice to randomize the response. *Commun. Stat. Simul. Comput.* **2017**, *46*, 4708–4721. [CrossRef]

© 2019 by the authors. Licensee MDPI, Basel, Switzerland. This article is an open access article distributed under the terms and conditions of the Creative Commons Attribution (CC BY) license (http://creativecommons.org/licenses/by/4.0/).

Article

Combination of Ensembles of Regularized Regression Models with Resampling-Based Lasso Feature Selection in High Dimensional Data

Abhijeet R Patil [1] and Sangjin Kim [2,*]

[1] Computational Science, University of Texas at El Paso, El Paso, TX 79968, USA; arpatil@miners.utep.edu
[2] Department of Mathematical Sciences, University of Texas at El Paso, El Paso, TX 79968, USA
* Correspondence: skim10@utep.edu

Received: 14 October 2019; Accepted: 4 January 2020; Published: 10 January 2020

Abstract: In high-dimensional data, the performances of various classifiers are largely dependent on the selection of important features. Most of the individual classifiers with the existing feature selection (FS) methods do not perform well for highly correlated data. Obtaining important features using the FS method and selecting the best performing classifier is a challenging task in high throughput data. In this article, we propose a combination of resampling-based least absolute shrinkage and selection operator (LASSO) feature selection (RLFS) and ensembles of regularized regression (ERRM) capable of dealing data with the high correlation structures. The ERRM boosts the prediction accuracy with the top-ranked features obtained from RLFS. The RLFS utilizes the lasso penalty with sure independence screening (SIS) condition to select the top k ranked features. The ERRM includes five individual penalty based classifiers: LASSO, adaptive LASSO (ALASSO), elastic net (ENET), smoothly clipped absolute deviations (SCAD), and minimax concave penalty (MCP). It was built on the idea of bagging and rank aggregation. Upon performing simulation studies and applying to smokers' cancer gene expression data, we demonstrated that the proposed combination of ERRM with RLFS achieved superior performance of accuracy and geometric mean.

Keywords: ensembles; feature selection; high-throughput; gene expression data; resampling; lasso; adaptive lasso; elastic net; SCAD; MCP

MSC: 62P10; 62F40; 62F07

1. Introduction

With the advances of high throughput technology in biomedical research, large volumes of high-dimensional data are being generated [1–3]. Some of the examples of what produces such data are microarray gene expression [4–6] data sequencing, RNA-seq [7], genome-wide association studies (GWASs) [8,9], and DNA-methylation studies [10,11]. These data are high dimensional in nature, where the total count of features is significantly larger than the number of samples ($p >> n$)—termed the curse of dimensionality. Although this is one of the major problems, there are many other problems, such as noise, redundancy, and over parameterization. To deal with these problems, many two-stage approaches of feature selection (FS) and classification algorithms have been proposed in machine learning over the last decade.

The FS methods are used to reduce the dimensionality of data by removing noisy and redundant features that help in selecting the truly important features. The FS methods are classified into rank-based and subset methods [12,13]. Rank-based methods rank all the features with respect to their importance based on some criteria. Although there is a lack of threshold to select the optimal number of top-ranked features, this can be solved using the sure independence screening (SIS) [14] conditions. Some of the popular rank-based FS methods used in bioinformatics are information gain [15], Fisher score [16], chi-square [17], and minimum redundancy maximum relevance [18]. These rank-based FS methods have several advantages, such as that they avoid overfitting and are computationally faster because they do not depend on the performances of classification algorithms. However, these methods do not consider joint importance because they focus on marginal significance. To overcome this issue, feature subset section methods were introduced. The subset methods [19] are the ones where the subsets of features are selected with some predetermined threshold based on some criteria, but these methods need more computational time in a high-dimensional data setting and lead to an NP-hard problem [20]. Some of the popular subset methods include Boruta [21] and relief [22].

For the classification of gene expression data, there are non-parametric-based popular algorithms, such as random forests [23], Adaboost [24], and support vector machines [25]. The support vector machines are known to perform well in highly correlated gene expression data compared to the random forests [26]. The random forests and Adaboost are based on the concept of decision trees, and the support vector machines are based on the idea of hyperplanes. In addition to the above, there are parametric machine learning algorithms, such as penalized logistic regression (PLR) models, that have five different penalties which are predominantly popular in high-dimensional data. The first two classifiers are Lasso [27] and ridge [28] that are based on L1 and L2 penalties. The third classifier is a combination of these and is termed as elastic net [29]. The other two PLR classifiers are SCAD [30] and MCP [31], which are based on non-concave and concave respectively. All these individual classifiers are very common in machine learning and bioinformatics [32]. However, in highly correlated gene expression data, the individual classifiers do not perform well in terms of prediction accuracy. To overcome the issue of individual classifiers, ensemble classifiers are proposed [33,34]. The ensemble classifiers are bagging and aggregating methods [35,36] that are employed to improve the accuracy of several "weak" classifiers [37]. The tree-based method of classification by ensembles from random partitions (CERP) [38] showed good performance but is computer-intensive. The ensembles of logistic regression models (LORENS) [39] for high-dimensional data were proven to be better for classification. However, there was a decrease in performance when there were a smaller number of true, important variables in the high-dimensional space because of random partitioning.

To address these issues, there is a need to develop a novel combination of FS with a classification method and compare the proposed method with the other combinations of popular FS with the classifiers through extensive simulation studies and a real data application. In a high dimensional data set, it is necessary to filter out the redundant and unimportant features using the FS methods. This helps in reducing the computational time and helps in boosting the prediction accuracy with the help of significant features.

In this article, we introduce the combination of an ensemble classifier with an FS method—the resampling-based lasso feature selection (RLFS) method for ranking features, and ensemble of regularized regression models (ERRM) for classification purposes. The resampling approach was proven to be one of the best FS screening steps in a high-dimensional data setting [13]. The RLFS uses the selection probability with lasso penalty, and the threshold for selecting the top-ranked features is set using b-SIS condition; and these select features were applied to the ERRM to achieve the best prediction accuracy. The ERRM uses five individual regularization models, lasso, adaptive lasso, elastic net, SCAD, and MCP.

2. Materials and Methods

The FS method includes the proposed RLFS method, information gain, chi-square, and minimum redundancy maximum relevance. The classification methods include support vector machines, penalized regression models, and tree-based methods, such as random forests and adaptive boosting. The programs for all the experiments were written using R software [40]. The FS and classification were performed with the packages [41–46] obtained from CRAN. The weighted rank aggregation was evaluated with the RankAggreg package obtained from [47]. The codes for implementing the algorithms are available at [48]. The SMK-CAN-187 data were obtained from [49]; some of the applications of the data can be found in the articles [50,51] where the importance of screening approach in high dimensional data is elaborated.

2.1. Data Setup

To assess the performances of the models, we developed simulation study and also considered a real application of gene expression data.

2.1.1. Simulation Data Setup

The data were generated based on a random multivariate normal distribution where the mean was assigned as 0, and the variance-covariance matrix Σ_x adapts a compound symmetry structure with the diagonal items set to 1 and the off-diagonal items being ρ values.

$$\Sigma_x = \begin{pmatrix} 1 & \rho & \cdots & \rho \\ \rho & 1 & \cdots & \rho \\ \vdots & \vdots & \ddots & \vdots \\ \rho & \rho & \cdots & 1 \end{pmatrix}_{p \times p}. \quad (1)$$

The class labels were generated using the Bernoulli trails with the following probability:

$$\pi_i(y_i = 1 | x_i) = \frac{\exp(x_i \beta)}{1 + \exp(x_i \beta)}. \quad (2)$$

The data matrix $x_i \sim N_p(0, \Sigma_x)$ was generated using the random multivariate normal distribution, and the response variable y_i was generated by binomial distribution, as shown in Equations (1) and (2) respectively. For sufficient comparison of the performance of the model and subsidizing the effects of the data splits, all of the regularized regression models were built using the 10-fold cross-validation procedure, and the averages were taken over 100 partitioning times referred to as 100 iterations in this paper. The data generated are high-dimensional in nature with the number of samples, $n = 200$ and total features, $p = 1000$. The true regression coefficients were set to 25, which were generated using uniform distribution with the minimum and maximum values 2 and 4, respectively.

With this setup of high-dimensional data, we simulated three different types of data, each with correlation structures $\rho = 0.2$, 0.5, and 0.8 respectively. These values show the low, intermediate, and high correlation structures in the datasets which are significantly similar to what we usually see in the gene expression or others among many types of data in the field of bioinformatics [13,52]. At first, the data were divided randomly into training and testing sets with 75% and 25% of samples respectively; 75% of the training data was given to the FS methods, which ranked the genes concerning their importance, and then the top-ranked genes were selected based on b-SIS condition. The selected genes were applied in all the classifiers. For standard comparison and mitigating the effects of the data splitting, all of the regularized regression models were built using the 10-fold cross-validation; the models were assessed for testing the performance with the testing data using different evaluation metrics, and averages were taken over 100 splitting times referred to as 100 iterations.

2.1.2. Experimental Data Setup

To test the performance of the proposed combination of ERRM with RLFS, and compare it with the rest of the combinations of FS and classifiers, the gene expression data SMK-CAN-187 were analyzed. The data include 187 samples and 19,993 genes obtained from smokers, which included 90 samples from those with lung cancer and 97 samples from those without lung cancer. This data is high-dimensional, with the number of genes being 19,993. The preprocessing procedures are necessary to handle these high-dimensional data. At first, the data were randomly divided into training and testing sets with 75% and 25% of samples respectively. As the first filtering step, 75% of the training data were given to the marginal maximum likelihood estimator (MMLE), to overcome the redundant noisy features, and the genes were ranked based on their level of significance. The ranked significant genes were next applied to the FS methods along with the proposed RLFS method as the second filtering step, and a final list of truly significant genes was obtained. These significant genes were applied to all the classification models along with the proposed ERRM classifier. All of the models were built using the 10-fold cross-validation. The average classification accuracy and Gmean of our proposed framework were tested using the test data. The above procedure was repeated for 100 times and the averages were taken.

2.1.3. Data Notations

Let the expression levels of features in ith sample be represented as $x_i = (x_{i1}, x_{i2}, ..., x_{ip})$ for $i = 1, ..., n$, where n is the total number of samples and p is the total number of features. The response variables, $y_i \in \{0, 1\}$, where $y_i = 0$ means that ith individual is in the non disease group and $y_i = 1$ is disease group.

The original data x_i were split into 75% for the training set x_j and 25% for the testing set x_k. The training set $x_j = (x_{j1}, x_{j2}, ..., x_{jp})$ for $j = 1, ..., t$, where t is the number of training samples, the response variable y_j for the training set. The testing set $x_k = (x_{k1}, x_{k2}, ..., x_{kp})$ for $k = 1, ..., v$, where v is the number of testing samples; the response variable is y_k for the testing set. The classifiers are fitted on x_j, and the class labels y_j as training data set to predict the classification of y_k using x_k of the testing set.

The detailed procedure is as follows. The training data x_j were given to the FS methods, and the new reduced feature set $x_r = (x_{j1}, x_{j2}, ..., x_{jf})$ for $j = 1, ..., t$, where t was the samples included in training data, and f was the reduced number of features after the FS step. This reduced feature set x_r was used as new training data for building the classification models.

2.2. Rank Based Feature Selection Methods

With the gain in popularity of high dimensional data in bioinformatics, the challenges to deal with it also grew. In gene expression data, having large p and small n problems, the n represents the samples as patients and p represents the features as genes. Dealing with such a large number of genes that are generated by conducting large biological experiments involves computationally intensive tasks that become too expensive to handle. The performance drops when such a large number of genes are added to the model. To overcome this problem, employing FS methods becomes a necessity. In statistical machine learning, there are many FS methods developed to deal with the gene expression data. But most of the existing algorithms are not completely robust applications to the gene expression data. Hence, we propose an FS method that ranks the features based on some criteria explained in the next section. We also explain some other popular FS methods in classification problems, such as information gain, chi-square, and minimum redundancy maximum relevance.

2.2.1. Information Gain

The information gain (IG) method [15] is simple, and one of the widely used FS methods. This univariate FS method is used to assess the quantity of information shared between the training feature set $x_j = (x_{j1}, x_{j2}, ..., x_{jp})$ for $j = 1, ..., t$, where t is the number of training samples, for $g = 1, 2, ...p$, where g is the feature in p number of features, and the response variable y_j. It provides an ordered ranking of all the features having a strong correlation with the response variable that helps to obtain good classification performance.

The information gain between the gth feature in x_j and the response variable y_j is given as follows:

$$IG(x_j; y_j) = H(x_j) - H(x_j|y_j), \qquad (3)$$

where $H(x_j)$ is entropy of x_j and $H(x_j|y_j)$ is entropy of x_j given y_j. The entropy [53] of x_j is defined by the following equation:

$$H(x_j) = \sum_{g \in x_j} \pi(g) \log(\pi(g)), \tag{4}$$

where g indicates discrete random variable x_j and $\pi(g)$ gives the probability of g on all values of x_j. Given the random variable y_j, the conditional entropy of x_j is:

$$H(x_j|y_j) = \sum_{y \in y_j} \pi(y) \sum_{g \in x_j} \pi(g|y) \log(\pi(g|y)), \tag{5}$$

where $\pi(y)$ is the prior probability of y_j; $\pi(g|y)$ is conditional probability of g in a given y that shows the uncertainty of x_j given y_j.

$$IG(x_j; y_j) = \sum_{g \in x_j} \sum_{y \in y_j} \pi(g, y) \log \frac{\pi(g, y)}{\pi(g)\pi(y)}, \tag{6}$$

where $\pi(g, y)$ is the joint probability of g and y. IG is symmetric such that $IG(x_j; y_j) = IG(y_j; x_j)$, and is zero if the variables x_j and y_j are independent.

2.2.2. Chi-Square Test

The chi-square test (Chi2) belongs to the category of the non-parametric test, which is used mainly in determining the significant relation between two categorical variables. As part of the preprocessing step, we used the "equal interval width" approach to transform the numerical variables into categorical counterparts. The "equal interval width" algorithm first divides the data into q intervals of equal size. The width of each interval is defined as: $w = (max - min)/q$ and the interval boundaries are determined by: $min + w, min + 2w, ..., min + (q-1)w$.

The general rule in Chi2 is that the features have a strong dependency on the class labels selected, and the features independent of the class labels are ignored.

From the training set, $x_j = (x_{j1}, ...x_{jp})$, $g = 1, 2, ...p$, where g is every feature in p number of features. Given a particular feature g with r different feature values [53], the Chi2 score of that particular feature can be calculated as:

$$\tilde{\chi}^2(g) = \sum_{j=1}^{r} \sum_{s=1}^{p} \frac{(O_{js} - E_{js})^2}{E_{js}}, \tag{7}$$

where O_{js} is the number of instances with the jth feature value given feature g. In addition, $E_{js} = \frac{O_{*s} O_{j*}}{O}$, where O_{j*} indicates the number of data instances with the feature value given feature g, O_{*s} denotes the number of data instances in r, and p is total number of features.

When two features are independent, the O_{js} is closer to the expected count E_{js}; consequently, we will have smaller Chi2 score. On the other hand, the higher Chi2 score implies that the feature is more dependent on the response and it can be selected for building the model during training.

2.2.3. Minimum Redundancy Maximum Relevance

The minimum redundancy and maximum relevance method (MRMR) is built on optimization criteria of mutual information (redundancy and relevance); hence, it is also defined under mutual information based methods. If a feature has uniformly of expressions or if they are randomly distributed in different classes, its mutual information with such classes is null [18]. If a feature is expressed differentially for different classes, it should have strong mutual information. Hence, we use mutual information as a measure of the relevance of features. MRMR also reduces the redundant features from the feature set. For a given set of features, it tries to measure both the redundancy among features and relevance between features and class vectors.

The redundancy and relevance are calculated based on mutual information, which is as follows: We know that, in the training set x_j, $g = 1, ..., p$ represents every feature in x_j and y_j is the response variable.

$$I(g,y) = \sum_{g \in x_j} \sum_{y \in y_j} \log \frac{\pi(g,y)}{\pi(g)\pi(y)}. \tag{8}$$

In the following equation, for simplicity, let us consider the training set x_j as X and response variable y_j as Y. The objective function is shown below:

$$J_{MRMR}(X_S, Y) = \frac{1}{|S|} \sum_{i \in S} I(X_i, Y) - \frac{1}{|S|^2} \sum_{i,j \in S} I(X_i, X_j), \tag{9}$$

where S is the subset of selected features and X_i is the ith feature. The first term is a measure of relevance that is the sum of mutual information of all the selected features in the set S with respect to the output Y. The second term is measure of redundancy that is the sum of the mutual information between all the selected features in the subset S. By optimizing the Equation (9), we are maximizing the first term and minimizing the second term simultaneously.

2.3. Classification Algorithms

Along with gene selection, improving prediction accuracy when dealing with high-dimensional data has always been a challenging task. There is a wide range of popular classification algorithms used when dealing with high throughput data, such as tree-based methods [54], support vector machines, and penalized regression models [55]. These popular models are discussed briefly in this section.

2.3.1. Logistic Regression

Logistic regression (LR) is perhaps one of the primary and popular models used while dealing with binary classification problems [56]. Logistic regression for dealing with more than two classes is called multinomial logistic regression. The primary focus here is on the binary classification. Given the set of inputs, the output is a predicted probability that the given input point belongs to a particular class. The output is always within [0, 1]. Logistic regression is based on the assumption that the original input space can be divided into two separate regions, one for each class, by a plane. This plane helps to discriminate between the dots belonging to different classes and is called as linear discriminant or linear boundary.

One of the limitations is the number of parameters that can be estimated needs to be smaller and should not exceed the number of samples.

2.3.2. Regularized Regression Models

Regularization is a technique used in logistic regression by employing penalties to overcome the limitations of dealing with high-dimensional data. Here, we discuss the PLR models such as lasso, adaptive lasso, elastic net, SCAD, and MCP. These five methods are included in the proposed ERRM and also tested as independent classifiers for comparing performance with the ERRM.

The logistic regression equation:

$$\log \left(\frac{\pi(y_j = 1|x_j)}{1 - \pi(y_j = 1|x_j)} \right) = \beta_0 + \beta x_j, \tag{10}$$

where $j = 1....t$ and $\beta = (\beta_1...\beta_f)^T$.

From logistic regression Equation (10), the log-likelihood estimator is shown as below:

$$l(\beta, y_j) = \sum_{j=1}^{t} \{y_j \log(\pi(y_j = 1|x_j)) + (1 - y_j)\log(1 - \pi(y_j = 1|x_j))\}. \quad (11)$$

Logistic regression offers the benefit by simultaneous estimation of the probabilities $\pi(x_j)$ and $1 - \pi(x_j)$ for each class. The criterion for prediction is $I\{\pi(y_j = 1|x_j) \geq 0.5\}$, where $I(\cdot)$ is an indicator function.

The parameters for PLR are estimated by minimizing above function:

$$\hat{\beta}_{PLR} = \underset{\beta}{\mathrm{argmin}} \left[-l(\beta, y_j) + p(\beta) \right], \quad (12)$$

where $p(\beta)$ is a penalty function, $l(\beta, y_j)$ is the log-likelihood function.

Lasso is a widely used method in variable selection and classification purposes in high dimensional data. It is one of the five methods used in the proposed ERRM for classification purposes. The LASSO penalized regression method is defined below:

$$\hat{\beta}_{LASSO} = \underset{\beta}{\mathrm{argmin}} \left[-l(\beta, y_j) + \lambda \sum_{j=1}^{f} |\beta_j| \right], \quad (13)$$

where f is the reduced number of features; λ is the tuning parameter that controls the strength of the L1 penalty.

The oracle property [30] has consistency in variable selection and asymptotic normality. The lasso works well in subset selection; however, it lacks the oracle property. To overcome this, different weights are assigned to different coefficients: this describes a weighted lasso called adaptive lasso. The adaptive lasso (ALASSO) penalty is shown below:

$$\hat{\beta}_{ALASSO} = \underset{\beta}{\mathrm{argmin}} \left[-l(\beta, y_j) + \lambda \sum_{j=1}^{f} w_j |\beta_j| \right], \quad (14)$$

where f is the reduced number of features, λ is the tuning parameter that controls the strength of the L2 penalty, and w_j is the weight vector based on ridge estimator. The ridge estimator [28] uses the L2 regularization method which obtains the size of coefficients by adding the L2 penalty.

The elastic net (ENET) [57] is the combination of lasso which uses the L1 penalty, and ridge which uses the L2 penalty. The sizable number of variables is obtained, which helps in avoiding the model turning into an excessively sparse model.

The ENET penalty is defined as:

$$\hat{\beta}_{ENET} = \underset{\beta}{\mathrm{argmin}} \left[-l(\beta, y_j) + \lambda \left(\frac{1-\alpha}{2} \sum_{j=1}^{f} |\beta_j|^2 + \alpha \sum_{j=1}^{f} |\beta_j| \right) \right], \quad (15)$$

where λ is the tuning parameter that controls the penalty, f is the number of features, α is the mixing parameter between ridge $\alpha = 0$ and lasso $\alpha = 1$.

The smoothly clipped absolute deviation penalty (SCAD) [30] is a sparse logistic regression model with a non-concave penalty function. It improves the properties of the L1 penalty. The regression coefficients are estimated by minimizing the log-likelihood function:

$$\hat{\beta}_{SCAD} = \underset{\beta}{\mathrm{argmin}} \left[-l(\beta, y_j) + \lambda \sum_{j=1}^{f} p_\lambda(\beta_j) \right]. \quad (16)$$

In Equation (16) the $p_\lambda(\beta_j)$ is defined by:

$$|\beta_i|I_{(|\beta_j|\leq\lambda)} + \left(\frac{\{(c^2-1)\lambda^2 - (c\lambda - |\beta_j|)_+^2\}I(\lambda \leq |\beta_j|)}{2(c-1)}\right), \quad c > 2 \text{ and } \lambda \geq 0. \tag{17}$$

Minimax concave penalty (MCP) [31] is very similar to the SCAD. However, the MCP relaxes the penalization rate immediately, while for SCAD, the rate remains smooth before it starts decreasing. The MCP equation is given as follows:

$$\hat{\beta}_{MCP} = \underset{\beta}{\operatorname{argmin}}\left[-l(\beta, y_j) + \lambda \sum_{j=1}^{f} p_\lambda(\beta_j)\right]. \tag{18}$$

In Equation (18) the $p_\lambda(\beta_j)$ is defined as:

$$\left(\frac{2c\lambda|\beta_j| - \beta_j^2}{2c}\right)I(|\beta_j| \leq c\lambda) + \left(\frac{c\lambda^2}{2}\right)I(|\beta_j| > c\lambda), \text{ for } \lambda \geq 0 \text{ and } c > 1. \tag{19}$$

2.3.3. Random Forests

The random forest (RF) [23] is an interpretive and straightforward method commonly used for classification purposes in bioinformatics. It is also known for its variable importance ranking in high dimensional data sets. RF is built on the concept of decision trees. Decision trees are usually more decipherable when dealing with binary responses. The idea of RF is to operate as an ensemble instead of relying on a single model. RF is a combination of a large number of decision trees where each tree has some random subset of features obtained from the data by allowing repetitions. This process is called bagging. The majority voting scheme is applied by aggregating all the tree models and obtaining one final prediction.

2.3.4. Support Vector Machines

Support vector machines (SVM) [25] are well known amongst most of the mainstream algorithms in supervised learning. The main goal of a SVM is to choose a hyperplane that can best divide the data in the high dimensional space. This helps to avoid overfitting. The SVM detects the maximum margin hyperplane, the hyperplane that maximizes the distance between the hyperplane, and the closest dots [58]. The maximum margin indicates that the classes are well separable and correctly classified. It is represented as a linear combination of training points. As a result, the decision boundary function for classifying points as to hyperplane only involves dot products between those points.

2.3.5. Adaboost

Adaboost is also known as adaptive boosting (AB) [24]. It improves the performance of a particular weak boosting classifier through an iterative process. This ensemble learning algorithm can be extensively applied to classification problems. The primary objective here is to assign more weights to the patterns that are harder to classify. Initially, the same weights are assigned to each training item. The weights of the wrongly classified items are incremented while the weights of the correctly classified items are decreased in each iteration. Hence, with the additional iterations and more classifiers, the weak learner is bound to cast on the challenging samples of the training set.

2.4. The Proposed Framework

We propose a combination of the FS method and classification method. For the filtering procedure, the resampling-based lasso feature selection method is introduced, and for the classification, the ensemble of regularized regression models is developed.

2.4.1. The Resampling-Based Lasso Feature Selection

From [13], we see that the resampling-based FS is relatively more efficient in comparison to the other existing FS methods in gene expression data. The RLFS method is based on the lasso penalized regression method and the resampling approach employed to obtain the ranked important features using the frequency.

The least absolute shrinkage and selection operator (LASSO) [27] estimator is based on L1-regularization. The L1-regularization method limits the size of coefficients pushes the unimportant regression coefficients to zero by using the L1 penalty. Due to this property, variable selection is achieved. It plays a crucial role in achieving better prediction accuracy along with the gene selection in bioinformatics.

$$\hat{\beta}_{lasso} = \underset{\beta}{\operatorname{argmin}} \left[-\sum_{j=1}^{t} \{y_j \log(\pi(y_j=1|x_j)) + (1-y_j)\log(1-\pi(y_j=1|x_j))\} + \lambda \sum_{j=1}^{p} |\beta_j| \right]. \quad (20)$$

The selection probability $S(f_m)$ of the features based on the lasso is shown in the below equation.

$$S(f_m) = \frac{1}{R} \sum_{i=1}^{R} \frac{1}{L} \sum_{j=1}^{L} I(\beta_{ijm} \neq 0), \text{ for } m = 1, 2, ..., p. \quad (21)$$

The b-SIS criteria to select the top k ranked features is defined by,

$$\left[b \times \frac{n}{\log(n)} \right], \quad (22)$$

where R is defined by the total number of resampling, L is total number of λ values, f_m is the feature indexed as i, p is total number of features, n is total number of samples, and β_{ijm} is defined as regression coefficient of mth feature and $I()$ indicator variable. Each R number of resamples and L number of values of λ are considered to build the variable selection model. The 10-fold cross validation is considered while building the model.

After ranking the features using the RLFS method, we employ the b-SIS approach to select the top features based on Equation (22) where b is set to two. The number of true important variables selected among the top b-SIS ranked features is calculated in each iteration and the average of this is taken over 100 iterations.

2.4.2. The Ensembles of Regularized Regression Models

LASSO, ALASSO, ENET, SCAD, and MCP are the five individual regularized regression models included as base learners in our ERRM. The role of bootstrapped aggregation or bagging is to reduce the variance by averaging over an "ensemble" of trees, which will improve the performance of weak classifiers. $B = B_1^k, ..., B_M^k$ is the number of random bootstrapped samples obtained from reduced training set x_r with corresponding class label y_j. The five regularized regression models are trained on each bootstrapped sample B named sub-training data, leading to $5 \times B$ models. These five regularized models are then trained using the 10-fold cross-validation to predict the classes on the out of bag samples called sub-testing data where the best model fit in each of the five regularized regression model is obtained. Henceforth, in each of the five regularized models, the best model is selected and the testing data x_k is applied to obtain the final list of predicted classes for each of these models. For binary classification problems, in addition to accuracy, the sensitivity and specificity are primarily sought. The E evaluation metrics are computed for each of these best models of five regularized models. In order to get an optimized classifier using all the evaluation measures E is essential, and this is achieved using weighted rank aggregation. Here, each of the regularized models is ranked based on the performance of E evaluation metrics. The models are ranked based on the increasing order of performance; in the case of a matching score of accuracy for two or more models, other metrics such as sensitivity and specificity are considered. The best performing model among the five models is obtained based on these ranks. This procedure is repeated to obtain the best performing model in each of the tree T. Finally, the majority voting procedure is applied over the T trees to obtain a final list of predicted classes. The test class label is applied to measure the final E measures for assessing the performance of the proposed ensembles. The Algorithm 1 defines the proposed ERRM procedure.

The complete workflow of the proposed RLFS-ERRM framework is shown in Figure 1.

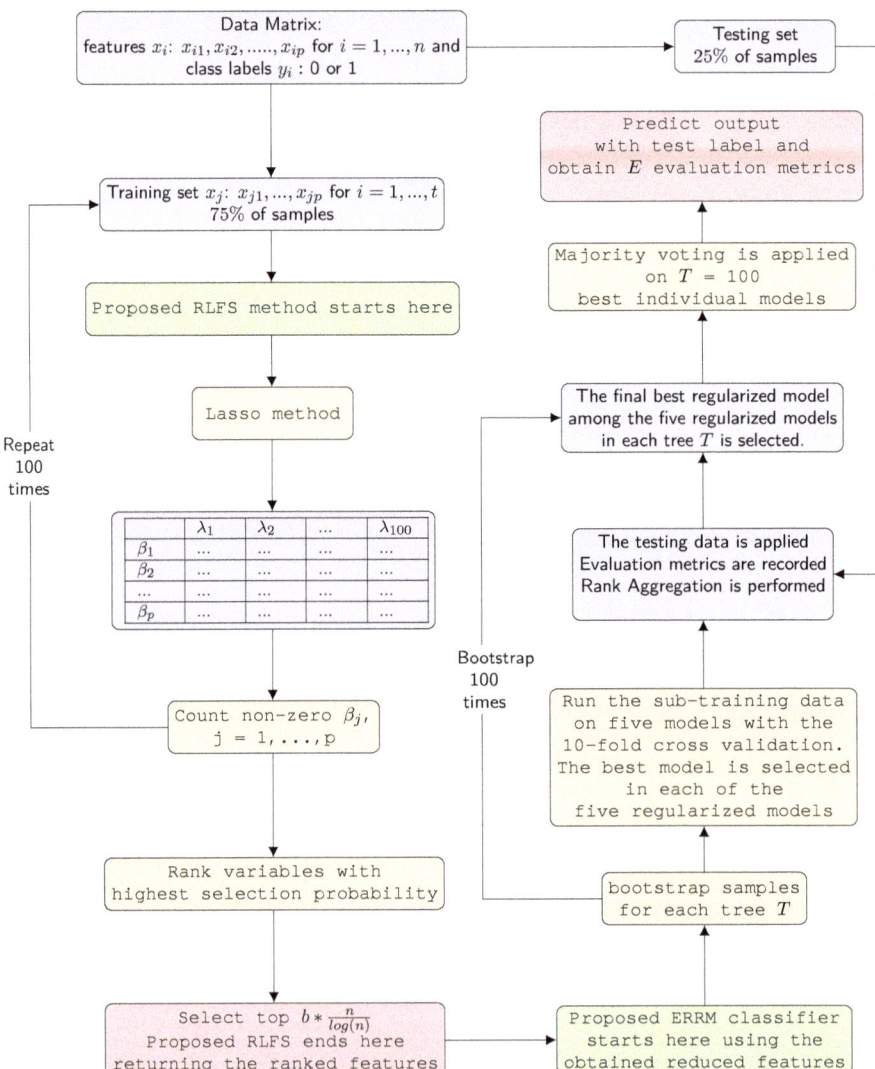

Figure 1. The complete workflow depicting the proposed combination of RLFS-ERRM framework.

Algorithm 1 Proposed ERRM
───
Step 1: Obtain new training data x_r with most informative features using the proposed RLFS method.

Step 2: Draw bootstrap samples from x_r and apply them to each of the regularized methods to be fitted with 10-fold cross validation.

Step 3: Apply out of bag samples (OOB) not used in bootstrap samples to the above fitted models to choose the best model using E evaluation metrics.

Step 4: Repeat steps 2 and 3 until getting 100 bootstrap models.

Step 5: Apply testing set x_k to each of 100 models to aggregate votes of classification.

Step 6: Predict classification of each sample by the rule of majority voting in the testing set.
───

2.5. Evaluation Metrics

We evaluated the results of combinations of FS methods with the classifier using accuracy and geometric mean (Gmean). The metrics are detailed with respect to true positive (TP), true negative (TN), false negative (FN), and false positive (FP). The equations for accuracy and Gmean are as follows:

$$\text{Accuracy} = \frac{TP + TN}{TP + TN + FP + FN} \qquad (23)$$

$$\text{Gmean} = \sqrt{\text{Sensitivity} \times \text{Specificity}},$$

where the sensitivity and specificity are given by:

$$\text{Sensitivity} = \frac{TP}{TP + FN} \quad \text{and} \quad \text{Specificity} = \frac{TN}{TN + FP}. \qquad (24)$$

3. Results

3.1. Simulation Results

The prediction performance of any given model is largely dependent on the type of the features. The features affecting the classification will help in attaining the best prediction accuracies. In Figure 2, we see the RLFS method with the top-ranked features based on the b-SIS criterion includes a higher number of true important features than other existing FS methods, such as IG, Chi2, and MRMR used for comparison in this study. The proposed RLFS performs consistently better across low, medium, and highly correlated simulated data, and the positive effect of having more true important variables was seen in all three simulation scenarios (further explained in detail).

3.1.1. Simulation Scenario (S1): Low Correlation 0.2

The predictors were generated, having a low correlation structure with $\rho = 0.2$. The proposed classifier ERRM performs better than the existing classifier on all the FS methods: proposed RLFS, IG, Chi2, and MRMR. Also, the proposed combination of ERRM classifier with RLFS method, with the accuracy and Gmean, each of which is 0.8606 and 0.8626 respectively, is relatively better in comparison to other combinations of FS method and classifier such as RLFS-LASSO, RLFS-ALASSO, RLFS-ENET, and the other remaining combinations, as observed in Figure 3. The combination of the FS method IG with proposed ERRM with an accuracy of 0.8476 is also seen performing better than IG-LASSO, IG-ALASSO, IG-ENET, IG-SCAD, IG-MCP, IG-AB, IG-RF, IG-LR, and IG-SVM. Similarly, the combination of Chi2-ERRM with an accuracy of 0.8538 is seen better than FS method Chi2 with the other remaining classifiers. The results are reported in Table 1. The combination of MRMR-ERRM has an accuracy of 0.8550 and Gmean of 0.8552 is better than the combination of FS method MRMR with the rest of the nine classifiers.

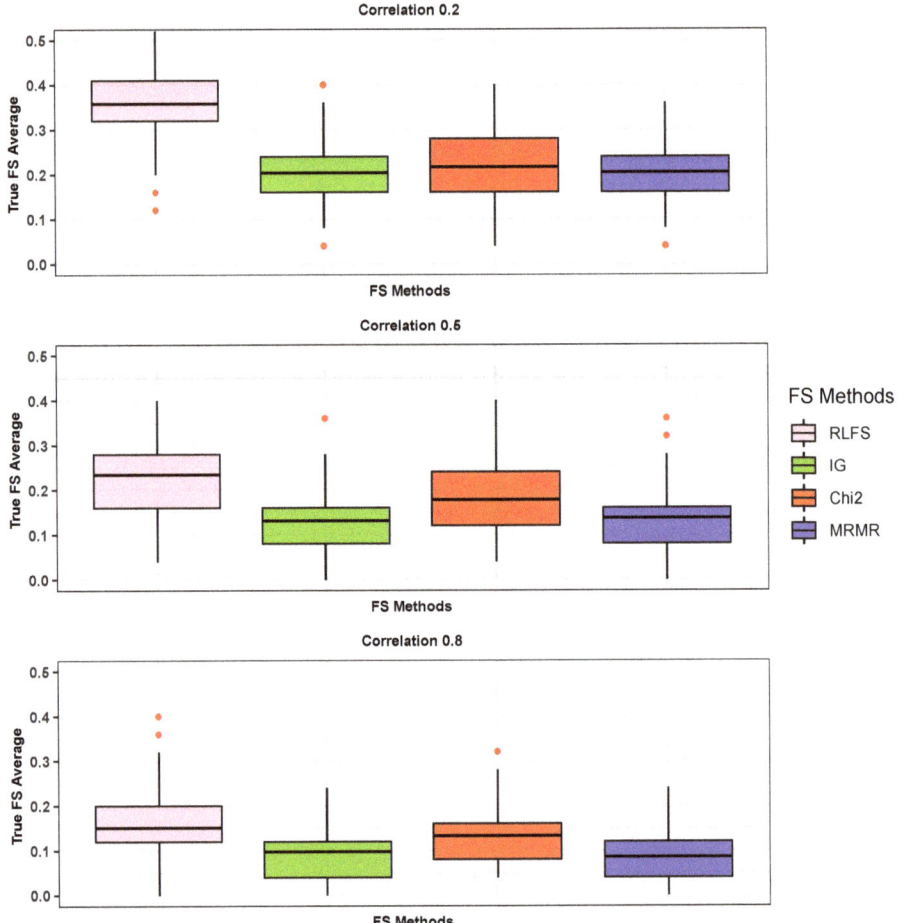

Figure 2. True number of features selected among top b-SIS ranked features, and the average of this taken over 100 iterations for three different scenarios. The first horizontal line in the box shows the first quartile and the second horizontal dark line which usually represents the median values are shown as the mean values in this article. The third horizontal line in each of the boxes shows the third quartile. The red dotted circles indicate the outliers in each of the FS methods.

All the classifiers with the features obtained from the RLFS method achieved the best accuracies in comparison to other FS methods, as seen in Figure 3. The combination of RLFS with SVM showed the second-best performance by attaining an accuracy of 0.8582, as seen in Table 1. The ENET method showed the best performance among all the regularized regression models with all the FS methods, and the best accuracy was obtained with the proposed RLFS method.

The proposed combination of RLFS-ERRM has better performance than the other existing combinations of the FS and classifier without the proposed FS method RLFS and classifier ERRM itself. For example, the existing FS methods IG, Chi2, and MRMR with the eight existing individual classifiers' performances are lower than the proposed RLFS-ERRM combination, as shown in Table 1.

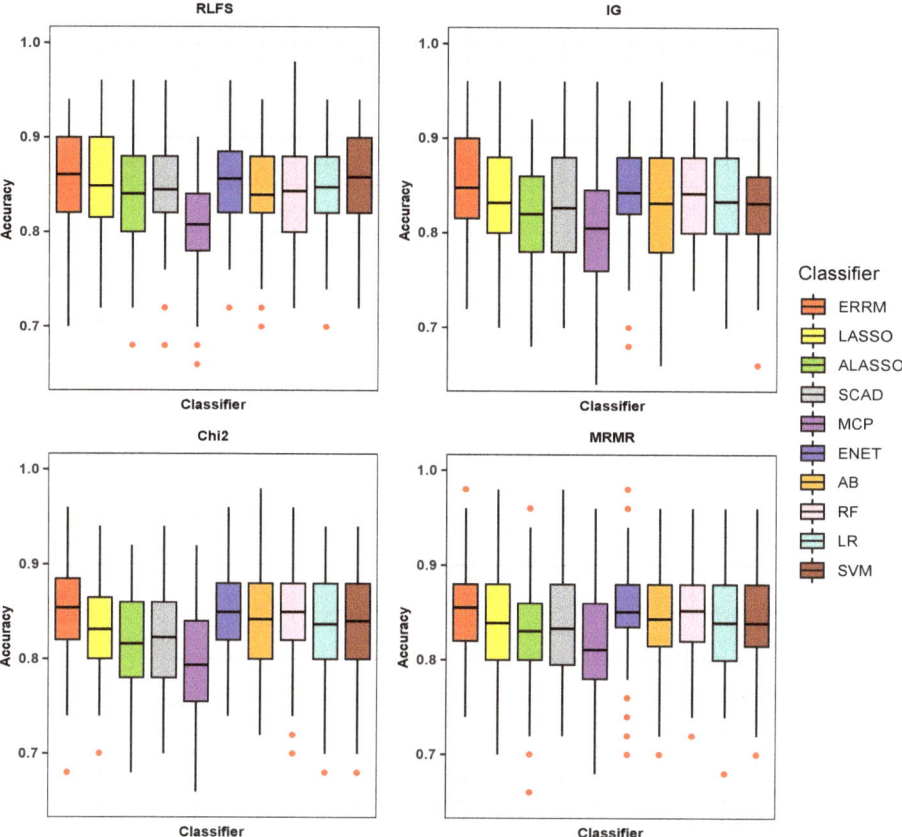

Figure 3. Comparison of accuracies of proposed combination of ensemble of regularized regression models (ERRM) with resampling-based lasso feature selection (RLFS) and other classifiers with feature selection methods when correlation = 0.2.

3.1.2. Simulation Scenario (S2): Intermediate Correlation 0.5

The predictor variables were generated using a medium correlation structure with $\rho = 0.5$. The proposed combination of the RLFS method and ERRM classifier, with the accuracy and Gmean, each of which is 0.9256 and 0.9266, respectively, attained relatively better performance compared to other combinations of the FS method and classifier such as RLFS-LASSO, RLFS-ALASSO, RLFS-ENET and the other remaining combinations. The results are shown in Table 1. From Figure 4, we see that the proposed ensemble classifier ERRM with other FS methods such as IG, Chi2, and MRMR performs best compared to the other nine individual classifiers.

The SVM and ENET classifiers with the RLFS method attained accuracies that are almost similar to the proposed combination of ERRM-RLFS. However, when Gmean is considered, the ERRM-RLFS outperforms the SVM combinations. The average SD of the proposed combination of the ERRM-RLFS is smaller than other combinations of the FS method and classifier. The accuracies of SVM and ENET classifiers with the IG method were 0.9128 and 0.9150 lower compared to the ERRM classifier with the IG method which had an accuracy of 0.9184. Similarly, the ERRM with the Chi2 method showed relatively better performance than the competitive classifiers ENET and SVM. Further, the ERRM classifier with the MRMR method having an accuracy of 0.9174 showed better performance than ENET, SVM, and other top-performing individual classifiers.

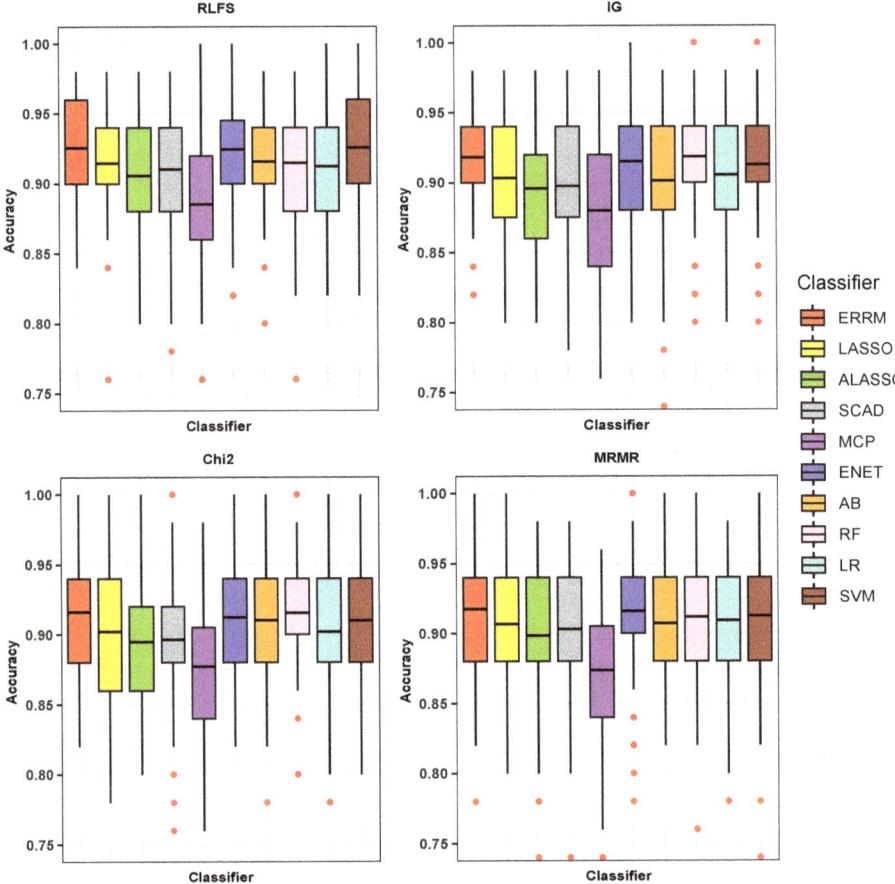

Figure 4. Boxplot showing the accuracies of classifiers with FS methods when correlation = 0.5.

While the SVM and ENET classifiers showed promising performance on the RLFS that had a good number of important features, they failed to show the same consistency on the other FS methods. On the other hand, the ensemble ERRM showed robust behavior, with being able to withstand the noise that helps in attaining better prediction accuracies and Gmean, not only with the RLFS method but also with other FS methods, such as IG, Chi2, and MRMR, as seen in Table 1.

Similar results are also found in the Simulation Scenario (S3): which has the highly correlated data with ρ set to 0.8. The results for this scenario are described in the Appendix A.

3.2. Experimental Results

Figure 5 shows the box plot of average accuracies taken over 100 iterations for all the combinations of FS and classifiers in experimental data. Each of the sub-figures in the figure shows the classifiers with the corresponding FS methods. As seen in Table 2, the performance of all the individual classifiers when applied on the RLFS method—the accuracy and Gmean—are relatively much better than the accuracies of the individual classifiers when applied on the IG, Chi2, and MRMR methods.

Table 1. Classification performance of proposed RLFS with ERRM compared to other combinations of feature selection methods with classifiers over 100 iterations.

Correlation	Classifier	Proposed RLFS		IG		Chi2		MRMR	
		Acc (SD)	Gmean (SD)	Accuracy (SD)	Gmean (SD)	Accuracy (SD)	Gmean (SD)	Accuracy (SD)	Gmean (SD)
0.2	Proposed ERRM	0.8606 (0.049)	0.8626 (0.073)	0.8476 (0.052)	0.8483 (0.079)	0.8538 (0.053)	0.8551 (0.071)	0.8550 (0.049)	0.8552 (0.075)
	LASSO	0.8486 (0.052)	0.8504 (0.075)	0.8316 (.054)	0.8335 (0.083)	0.8310 (0.052)	0.8323 (0.071)	0.8388 (0.051)	0.8393 (0.077)
	ALASSO	0.8402 (0.054)	0.8416 (0.077)	0.8198 (0.051)	0.8217 (0.079)	0.8160 (0.053)	0.8171 (0.075)	0.8304 (0.051)	0.8313 (0.079)
	ENET	0.8564 (0.048)	0.8584 (0.072)	0.8424 (0.054)	0.8441 (0.081)	0.8494 (0.046)	0.8509 (0.067)	0.8508 (0.052)	0.8508 (0.077)
	SCAD	0.8440 (0.054)	0.8457 (0.080)	0.8264 (0.057)	0.8283 (0.086)	0.8226 (0.061)	0.8239 (0.077)	0.8330 (0.056)	0.8336 (0.081)
	MCP	0.8078 (0.049)	0.8095 (0.081)	0.8050 (0.062)	0.8074 (0.088)	0.7936 (0.060)	0.7952 (0.085)	0.8110 (0.060)	0.8126 (0.082)
	AB	0.8390 (0.051)	0.8224 (0.077)	0.8314 (0.060)	0.8328 (0.080)	0.8422 (0.054)	0.8435 (0.075)	0.8432 (0.054)	0.8437 (0.075)
	RF	0.8432 (0.057)	0.8467 (0.084)	0.8414 (0.052)	0.8435 (0.078)	0.8498 (0.053)	0.8520 (0.075)	0.8522 (0.051)	0.8534 (0.077)
	LR	0.8474 (0.050)	0.8489 (0.076)	0.8330 (0.053)	0.8346 (0.080)	0.8370 (0.054)	0.8380 (0.073)	0.8394 (0.051)	0.8394 (0.080)
	SVM	0.8582 (0.049)	0.8595 (0.070)	0.8312 (0.052)	0.8320 (0.083)	0.8404 (0.054)	0.8416 (0.074)	0.8388 (0.049)	0.8378 (0.084)
0.5	Proposed ERRM	0.9256 (0.037)	0.9266 (0.053)	0.9184 (0.039)	0.9195 (0.059)	0.9160 (0.038)	0.9165 (0.056)	0.9174 (0.042)	0.9176 (0.056)
	LASSO	0.9146 (0.037)	0.9155 (0.053)	0.9034 (0.045)	0.9046 (0.061)	0.9020 (0.043)	0.9029 (0.063)	0.9066 (0.045)	0.9065 (0.062)
	ALASSO	0.9056 (0.039)	0.9062 (0.056)	0.8956 (0.044)	0.8966 (0.065)	0.8948 (0.046)	0.8954 (0.065)	0.8984 (0.046)	0.8982 (0.062)
	ENET	0.9244 (0.038)	0.9253 (0.052)	0.9150 (0.044)	0.9163 (0.061)	0.9122 (0.039)	0.9130 (0.060)	0.9158 (0.043)	0.9155 (0.058)
	SCAD	0.9102 (0.041)	0.9110 (0.060)	0.8974 (0.046)	0.8986 (0.063)	0.8964 (0.045)	0.8972 (0.065)	0.9090 (0.045)	0.9090 (0.059)
	MCP	0.8850 (0.047)	0.8855 (0.066)	0.8798 (0.050)	0.8813 (0.065)	0.8772 (0.045)	0.8782 (0.065)	0.8738 (0.049)	0.8738 (0.070)
	AB	0.9158 (0.035)	0.9166 (0.050)	0.9014 (0.046)	0.9027 (0.065)	0.9102 (0.040)	0.9112 (0.060)	0.9072 (0.047)	0.9075 (0.062)
	RF	0.9148 (0.039)	0.9166 (0.055)	0.9186 (0.041)	0.9199 (0.059)	0.9154 (0.042)	0.9167 (0.060)	0.9116 (0.043)	0.9127 (0.060)
	LR	0.9124 (0.037)	0.9127 (0.054)	0.9054 (0.043)	0.9063 (0.061)	0.9018 (0.045)	0.9024 (0.063)	0.9092 (0.043)	0.9084 (0.060)
	SVM	0.9256 (0.038)	0.9261 (0.054)	0.9128 (0.038)	0.9135 (0.056)	0.9080 (0.043)	0.9099 (0.061)	0.9126 (0.045)	0.9120 (0.062)

When we look at the performances of all the classifiers with the IG method in comparison to other FS methods, there is much variation in the accuracies, as seen in Figure 5. The SVM classifier, which attained the accuracy of 0.7026 with the RLFS method, dropped to 0.6422 with the IG method.

The proposed combination of the ERRM classifier with the RLFS method achieved the highest average accuracy of 0.7161, and the Gmean of 0.7127 outperformed the rest of the combinations of classifier with the FS method. The RLFS method is also a top-performing FS method on all individual classifiers. However, among the other FS methods, the MRMR method, when applied to all the individual classifiers, showed relatively much better performance than the application of IG and Chi2 methods to the individual classifiers. The second best-performing method is the ENET-RLFS combination, which had an accuracy of 0.7138. The SVM-IG combination showed the lowest performance with an accuracy of 0.6422 among all the combinations of the classifier with FS methods, as shown in Table 2.

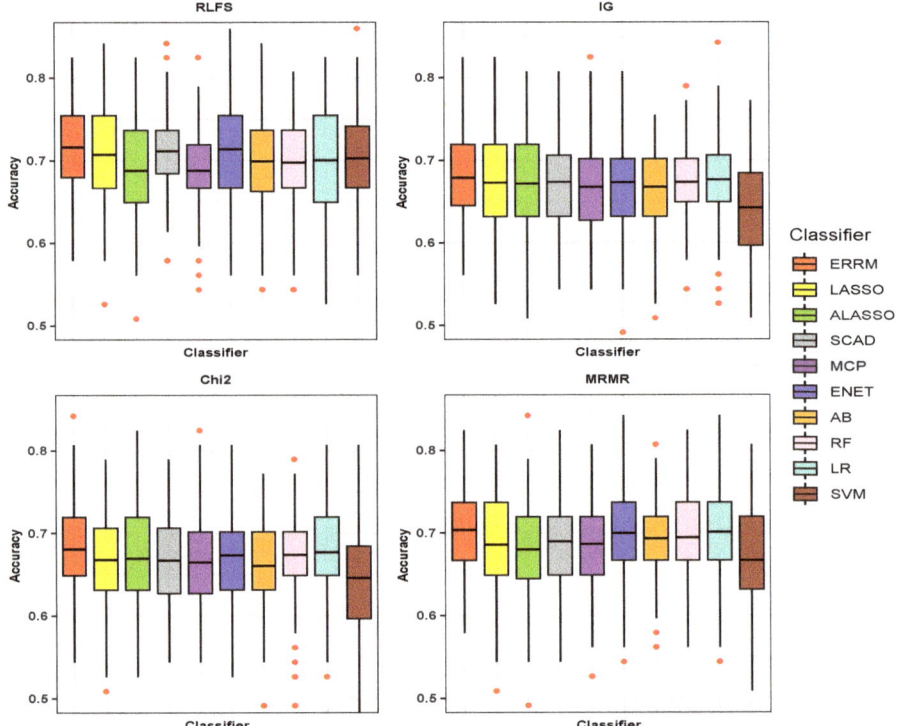

Figure 5. Boxplot showing the accuracies of classifiers with FS methods in experimental data SMK-CAN-187.

Table 2. Average values taken over 100 iterations in experimental data: SMK-CAN-187.

Classifier	Proposed RLFS		IG		Chi2		MRMR	
	Accuracy (SD)	Gmean (SD)	Accuracy (SD)	Gmean (SD)	Accuracy (SD)	Gmean (SD)	Accuracy (SD)	Gmean (SD)
Proposed ERRM	0.7161 (0.053)	0.7127 (0.082)	0.6789 (0.056)	0.6791 (0.091)	0.6807 (0.056)	0.6808 (0.091)	0.7035 (0.056)	0.7024 (0.087)
LASSO	0.7073 (0.064)	0.7058 (0.087)	0.6726 (.060)	0.6725 (0.060)	0.6680 (0.057)	0.6680 (0.090)	06859 (0.061)	0.6871 (0.097)
ALASSO	0.6878 (0.065)	0.6869 (0.091)	0.6715 (0.060)	0.6714 (0.094)	0.6696 (0.064)	0.6698 (0.092)	0.6800 (0.059)	0.6803 (0.092)
ENET	0.7138 (0.061)	0.7116 (0.085)	0.6733 (0.057)	0.6722 (0.093)	0.6733 (0.052)	0.6726 (0.090)	0.6998 (0.061)	0.6992 (0.095)
SCAD	0.7114 (0.054)	0.7098 (0.083)	0.6735 (0.056)	0.6732 (0.090)	0.6670 (0.058)	0.6669 (0.091)	0.6894 (0.059)	0.6901 (0.091)
MCP	0.6880 (0.010)	0.6870 (0.082)	0.6673 (0.057)	0.6663 (0.089)	0.6647 (0.059)	0.6639 (0.092)	0.6866 (0.057)	0.6874 (0.089)
AB	0.6991 (0.064)	0.6958 (0.087)	0.6673 (0.054)	0.6634 (0.086)	0.6605 (0.058)	0.6583 (0.094)	0.6929 (0.050)	0.6897 (0.083)
RF	0.6975 (0.056)	0.6933 (0.089)	0.6729 (0.045)	0.6691 (0.078)	0.6738 (0.054)	0.6703 (0.090)	0.6942 (0.055)	0.6902 (0.088)
LR	0.7001 (0.065)	0.6987 (0.089)	0.6761 (0.058)	0.6662 (0.097)	0.6770 (0.059)	0.6769 (0.094)	0.7008 (0.058)	0.7000 (0.086)
SVM	0.7026 (0.058)	0.7014 (0.086)	0.6422 (0.059)	0.6430 (0.099)	0.6459 (0.066)	0.6477 (0.105)	0.6668 (0.058)	0.6658 (0.092)

For assessing the importance of bootstrapping and FS screening of the proposed framework, we measured the performance of ERRM without FS screening. The results in Table 3 shows the results of ensembles method with and without bootstrapping procedure. We see that having the bootstrapping approach which is random sampling with replacement is a better approach in the ensembles.

Table 3. Comparison of proposed ERRM with and without bootstrapping.

	Bootstrapping	Accuracy (SD)	Gmean (SD)
ERRM without feature selection (FS) screening	Yes	0.7129 (0.053)	0.7093 (0.091)
ERRM without FS screening	No	0.6947 (0.057)	0.6944 (0.089)

The performance of the regularized regression models used in the proposed ensembles algorithm is tested with the FS screening method and without the FS screening method. In the former approach, the regularized regression models were built and tested using the proposed RLFS screening method with the selected amount of significant features, whereas in the latter approach, the regularized models used all the features for building the model. The performances of the penalized models with the FS screening showed better accuracies and Gmean than without FS screening, as reported in Table 4.

Table 4. Comparison of regularized regression models used in the ERRM with and without FS screening.

	FS Screening	Accuracy (SD)	Gmean (SD)
LASSO	Yes	0.7073 (0.064)	0.7058 (0.087)
	No	0.6740 (0.061)	0.6752 (0.125)
ALASSO	Yes	0.6878 (0.065)	0.6869 (0.091)
	No	0.6740 (0.061)	0.6752 (0.125)
ENET	Yes	0.7138 (0.061)	0.7116 (0.085)
	No	0.6740 (0.061)	0.6752 (0.125)
SCAD	Yes	0.7114 (0.054)	0.7098 (0.083)
	No	0.6740 (0.061)	0.6752 (0.125)
MCP	Yes	0.6880 (0.010)	0.6870 (0.082)
	No	0.6740 (0.061)	0.6752 (0.125)

4. Discussion

We investigated the performance of the proposed combination of ERRM with the RLFS method using simulation studies and a real data application. The RLFS method ranks the features by employing the lasso method with a resampling approach and the b-SIS criteria to set the threshold for selecting the optimal number of features, and these features are applied on the ERRM classifier, which uses bootstrapping and rank aggregation to select the best performing model across the bootstrapped samples and helps in attaining the best prediction accuracy in a high dimensional setting. The ensemble framework ERRM was built using five different regularized regression models. The regularized regression models are known for having the best performances in terms of variable selection and prediction accuracy on gene expression data.

To show the performance of our proposed framework, we used three different simulation scenarios with low, medium, and high correlation structures that matched the gene expression data. To further illustrate our point, we also used SMK-CAN-187 data. Figure 2 shows the boxplots of the average number of true important features, where the RLFS shows higher detection power than the other FS methods such as IG, Chi2, and MRMR. From the results of both simulation studies and experimental data, we showed that all the individual classifiers with the RLFS method performed much better compared to the IG, Chi2, and MRMR. We also observed that all the individual classifiers showed much instability with the other three FS methods. This means that the individual classifiers do not work well with more noise and less true important variables in the model. The SVM and ENET classifiers with all the FS methods performed a little better among all the classifiers. However, the performance was relatively still low in comparison to the proposed ERRM classifier with every FS method. The tree-based ensemble methods RF and AB with RLFS also attained good accuracies but were not the best compared to the ERRM classifier.

The proposed ERRM method was assessed with the FS screening and without the FS screening step along with the bootstrapping option. The ERRM with FS screening and bootstrapping approach works better than ERRM without the FS screening and bootstrapping technique. Also, the results from Table 3 show that the ensemble with bootstrapping is a better approach to both the filtered and unfiltered data. On comparing the performance of the individual regularized regression models used in the ensembles, the individual models with the proposed RLFS screening step showed comparatively better accuracy in comparison to the individual regularized regression models without the FS screening. This means that using the reduced number of significant features with RLFS is a better approach instead of using all the features from the data.

The importance of FS method was not addressed in any of the ensemble approaches [37–39], and the classification accuracies achieved by the corresponding proposed methods were much closer to the accuracies attained by existing approaches. In this paper, we compared the various combination of FS methods with different classifiers. The ERRM showed better overall performance not only with the RLFS but also with the other FS methods compared in this study. This means that the ERRM is robust and works much better on the highly correlated gene expression data. The rule of thumb fpr attaining the best prediction accuracy is that more the true important variables, better the prediction accuracy. Henceforth, from the results of simulation and experimental data, we see that the proposed combination of RLFS-ERRM is better compared to the other existing combinations of FS and classifiers, as seen in the Tables 1 and 2. The proposed ERRM classifier showed the best performance across all the FS methods, with the highest performance achieved with the RLFS method. The proposed RLFS method attained a higher number of significant features compared to other FS methods. However, the drawback is that with the increase in the correlation structure, there is a decreasing performance in selecting the significant features, as shown in Figure 2. The ensembles algorithms are known to be computationally expensive [39] because of the tree-based nature. However, in our proposed framework, before the ensembles of ERRM, we apply FS methods to remove the irrelevant features and keep significant features. This filtering step not only helps with improving prediction accuracy but also with overcoming the drawback of computational time required, as the number of features processed becomes lower.

5. Conclusions

In this paper, we proposed a combination of the ensembles of regularized regression models (ERRM) with resampling-based lasso feature selection (RLFS) for attaining better prediction accuracies in high dimensional data. We conducted extensive simulation studies where we showed the better performance of RLFS in detecting the significant features than other competitive FS methods. The ensemble classifier ERRM also showed better average prediction accuracy with the RLFS, IG, Chi2, and MRMR compared to other classifiers with these FS methods. We also saw an improved performance in the ensemble method when used with bootstrapping. On comparing the performances of individual regularized regression models, all the models showed an increase in their accuracies with the FS screening approach. In both the simulation study and the experimental data SMK-CAN-187, the better performance was achieved by the proposed combination of RLFS and ERRM compared to all other combinations of FS and classifiers. The minor drawback in the proposed framework is that, in the case of highly correlated data, there is smaller number of significant features selected with all the FS methods. As future work, we plan to focus on improving the detecting power of true important genes with the new FS method.

Author Contributions: Conceptualization, S.K. and A.R.P.; formal analysis and investigation, A.R.P.; Writing—Original draft preparation, A.R.P.; supervision and reviewing, S.K. All authors have read and agreed to the published version of the manuscript.

Funding: This research received no external funding.

Acknowledgments: We would like to thank the editors and three anonymous reviewers for their insightful comments and helpful suggestions.

Conflicts of Interest: The authors declare no conflict of interest.

Abbreviations

The following abbreviations are used in this manuscript:

FS	feature Selection
RLFS	resampling-based lasso feature selection
ERRM	ensemble regularized regression models
IG	information gain
Chi2	chi-square
MRMR	minimum redundancy maximum relevance
ALASSO	adaptive lasso
AB	adaptive boosting
RF	random forests
LR	logistic regression
SVM	support vector machines
SD	standard deviation

Appendix A

The data are generated based on a high correlation data structure with $\rho = 0.8$. The performance of the proposed combination of RLFS-ERRM is relatively better than the other combinations of the FS methods and classifiers. The results for simulation scenario S3 are shown in Figure A1. The average accuracies and Gmeans for all the FS and classifiers are noted in Table A1. The SVM and ENET classifiers with all the FS methods showed a little better performance among all individual classifiers. However, the accuracies and Gmeans attained by the proposed ensemble classifier ERRM with the FS methods RLFS, IG, and Chi2 were relatively better compared to the individual classifiers with FS methods. The best performance was achieved by the proposed RLFS-ERRM combination with an accuracy of 0.9586 and Gmean of 0.9596. The second-best performing combination was MRMR-SVM. The lowest performance in terms of accuracy and the Gmean was shown by Chi2-MCP. The MCP classifier has the lowest accuracy with all the FS methods. This explains why the MCP does not perform well when the predictor variables are highly correlated.

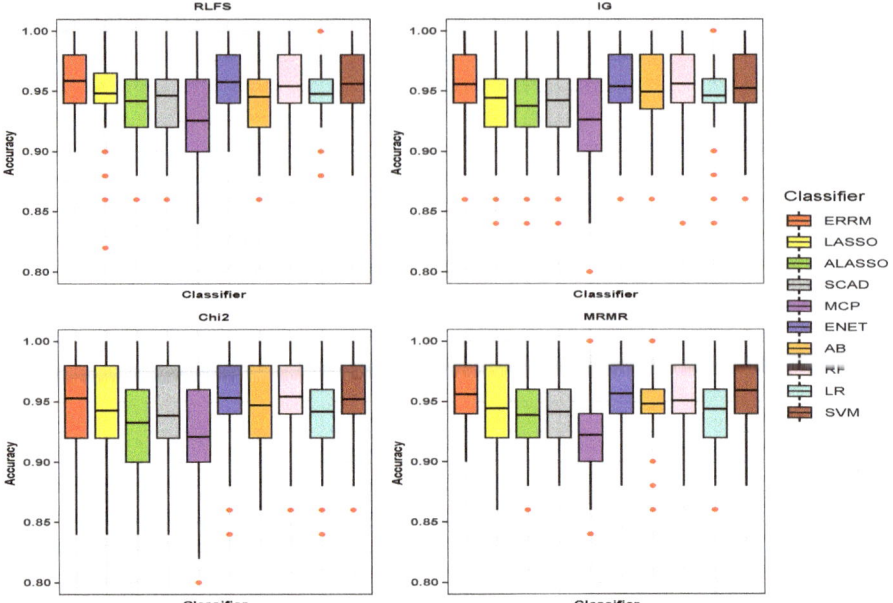

Figure A1. Boxplot showing the accuracies of Classifiers with FS methods in simulation scenario: S3 (Correlation = 0.8).

Table A1. Average values taken over 100 iterations in simulation scenario: S3 (High correlation: 0.8).

Classifier	Proposed RLFS		IG		Chi2		MRMR	
	Accuracy (SD)	Gmean (SD)	Accuracy (SD)	Gmean (SD)	Accuracy (SD)	Gmean (SD)	Accuracy (SD)	Gmean (SD)
Proposed ERRM	0.9586 (0.025)	0.9596 (0.039)	0.9556 (0.027)	0.9565 (0.041)	0.9530 (0.034)	0.9544 (0.045)	0.9560 (0.024)	0.9558 (0.037)
LASSO	0.9482 (0.033)	0.9493 (0.050)	0.9442 (.030)	0.9194 (0.045)	0.9428 (0.037)	0.9447 (0.051)	0.9444 (0.032)	0.9442 (0.042)
ALASSO	0.9420 (0.031)	0.9425 (0.051)	0.9376 (0.030)	0.9379 (0.045)	0.9328 (0.041)	0.942 (0.056)	0.9388 (0.033)	0.9389 (0.047)
ENET	0.9576 (0.025)	0.9587 (0.039)	0.9538 (0.029)	0.9546 (0.042)	0.9532 (0.034)	0.9546 (0.045)	0.9566 (0.024)	0.9562 (0.036)
SCAD	0.9464 (0.031)	0.9475 (0.049)	0.9422 (0.030)	0.9428 (0.045)	0.9386 (0.043)	0.9401 (0.055)	0.9414 (0.031)	0.9408 (0.043)
MCP	0.9256 (0.040)	0.9270 (0.062)	0.9262 (0.038)	0.9269 (0.055)	0.9210 (0.041)	0.9221 (0.058)	0.9224 (0.034)	0.9223 (0.048)
AB	0.9454 (0.032)	0.9469 (0.047)	0.9494 (0.030)	0.9501 (0.044)	0.9470 (0.034)	0.9482 (0.046)	0.9480 (0.029)	0.9481 (0.040)
RF	0.9540 (0.030)	0.9557 (0.043)	0.9560 (0.029)	0.9565 (0.043)	0.9542 (0.032)	0.9556 (0.044)	0.9508 (0.027)	0.9510 (0.039)
LR	0.9478 (0.029)	0.9482 (0.045)	0.9462 (0.030)	0.9469 (0.044)	0.9418 (0.038)	0.9432 (0.050)	0.9438 (0.028)	0.9437 (0.041)
SVM	0.9560 (0.027)	0.9568 (0.041)	0.9522 (0.030)	0.9527 (0.043)	0.9520 (0.031)	0.9526 (0.042)	0.9594 (0.026)	0.9587 (0.037)

References

1. Tariq, H.; Eldridge, E.; Welch, I. An efficient approach for feature construction of high-dimensional microarray data by random projections. *PLoS ONE* **2018**, *13*, e0196385. [CrossRef] [PubMed]
2. Bhola, A.; Singh, S. Gene Selection Using High Dimensional Gene Expression Data: An Appraisal. *Curr. Bioinform.* **2018**, *13*, 225–233. [CrossRef]
3. Dai, J.J.; Lieu, L.H.; Rocke, D.M. Dimension reduction for classification with gene expression microarray data. *Stat. Appl. Genet. Mol. Biol.* **2006**, *5*, 6. [CrossRef] [PubMed]
4. Lu, J.; Kerns, R.T.; Peddada, S.D.; Bushel, P.R. Principal component analysis-based filtering improves detection for Affymetrix gene expression arrays. *Nucleic Acids Res.* **2011**, *39*, e86. [CrossRef]
5. Bourgon, R.; Gentleman, R.; Huber, W. Reply to Talloen et al.: Independent filtering is a generic approach that needs domain specific adaptation. *Proc. Natl. Acad. Sci. USA* **2010**, *107*, E175–E175. [CrossRef]
6. Bourgon, R.; Gentleman, R.; Huber, W. Independent filtering increases detection power for high-throughput experiments. *Proc. Natl. Acad. Sci. USA* **2010**, *107*, 9546–9551. [CrossRef]
7. Ramsköld, D.; Wang, E.T.; Burge, C.B.; Sandberg, R. An Abundance of Ubiquitously Expressed Genes Revealed by Tissue Transcriptome Sequence Data. *PLoS Comput. Biol.* **2009**, *5*, e1000598. [CrossRef]
8. Li, L.; Kabesch, M.; Bouzigon, E.; Demenais, F.; Farrall, M.; Moffatt, M.; Lin, X.; Liang, L. Using eQTL weights to improve power for genome-wide association studies: A genetic study of childhood asthma. *Front. Genet.* **2013**, *4*, 103. [CrossRef]
9. Calle, M.L.; Urrea, V.; Vellalta, G.; Malats, N.; Steen, K.V. Improving strategies for detecting genetic patterns of disease susceptibility in association studies. *Stat. Med.* **2008**, *27*, 6532–6546. [CrossRef]
10. Bock, C. Analysing and interpreting DNA methylation data. *Nat. Rev. Genet.* **2012**, *13*, 705–719. [CrossRef]
11. Sun, H.; Wang, S. Penalized logistic regression for high-dimensional DNA methylation data with case-control studies. *Bioinformatics* **2012**, *28*, 1368–1375. [CrossRef]
12. Kim, S.; Halabi, S. High Dimensional Variable Selection with Error Control. *BioMed Res. Int.* **2016**, *2016*, 8209453. [CrossRef] [PubMed]
13. Kim, S.; Kim, J.M. Two-Stage Classification with SIS Using a New Filter Ranking Method in High Throughput Data. *Mathematics* **2019**, *7*, 493. [CrossRef]
14. Fan, J.; Lv, J. Sure independence screening for ultrahigh dimensional feature space. *J. R. Stat. Soc. Ser. B* **2008**, *70*, 849–911. [CrossRef] [PubMed]
15. Quinlan, J.R. *C4.5: Programs for Machine Learning*; Morgan Kaufmann Publishers Inc.: San Francisco, CA, USA, 1993.
16. Okeh, U.; Oyeka, I. Estimating the Fisher's Scoring Matrix Formula from Logistic Model. *Am. J. Theor. Appl. Stat.* **2013**, *2013*, 221–227.
17. Guyon, I.; Elisseeff, A. An Introduction to Variable and Feature Selection. *J. Mach. Learn. Res.* **2003**, *3*, 1157–1182.
18. Peng, H.; Long, F.; Ding, C. Feature Selection Based on Mutual Information: Criteria of Max-Dependency, Max-Relevance, and Min-Redundancy. *IEEE Trans. Pattern Anal. Mach. Intell.* **2005**, *27*, 1226–1238. [CrossRef]
19. Ditzler, G.; Morrison, J.C.; Lan, Y.; Rosen, G.L. Fizzy: Feature subset selection for metagenomics. *BMC Bioinform.* **2015**, *16*, 358. [CrossRef]
20. Su, C.T.; Yang, C.H. Feature selection for the SVM: An application to hypertension diagnosis. *Expert Syst. Appl.* **2008**, *34*, 754–763. [CrossRef]
21. Kursa, M.B.; Rudnicki, W.R. Feature Selection with the Boruta Package. 2010.
22. Urbanowicz, R.J.; Meeker, M.; Cava, W.L.; Olson, R.S.; Moore, J.H. Relief-based feature selection: Introduction and review. *J. Biomed. Inform.* **2017**, *85*, 189–203. [CrossRef] [PubMed]
23. Breiman, L. Random Forests. *Mach. Learn.* **2001**, *45*, 5–32. [CrossRef]
24. Freund, Y. An Adaptive Version of the Boost by Majority Algorithm. *Mach. Learn.* **2001**, *43*, 293–318. [CrossRef]
25. Hearst, M.A.; Dumais, S.T.; Osuna, E.; Platt, J.; Scholkopf, B. Support vector machines. *IEEE Intell. Syst. Appl.* **1998**, *13*, 18–28. [CrossRef]
26. Statnikov, A.R.; Wang, L.; Aliferis, C.F. A comprehensive comparison of random forests and support vector machines for microarray-based cancer classification. *BMC Bioinform.* **2008**, *9*, 319–319. [CrossRef] [PubMed]
27. Tibshirani, R. Regression Shrinkage and Selection via the Lasso. *J. R. Stat. Soc. Ser. B (Methodol.)* **1996**, *58*, 267–288. [CrossRef]
28. Marquardt, D.W.; Snee, R.D. Ridge Regression in Practice. *Am. Stat.* **1975**, *29*, 3–20.

29. Yang, X.G.; Lu, Y. Informative Gene Selection for Microarray Classification via Adaptive Elastic Net with Conditional Mutual Information. *arXiv* **2018**, arXiv:1806.01466.
30. Fan, J.; Li, R. Variable Selection via Nonconcave Penalized Likelihood and its Oracle Properties. *J. Am. Stat. Assoc.* **2001**, *96*, 1348–1360. [CrossRef]
31. Zhang, C.H. Nearly unbiased variable selection under minimax concave penalty. *Ann. Stat.* **2010**, *38*, 894–942. [CrossRef]
32. Hastie, T.; Tibshirani, R.; Friedman, J. *The Elements of Statistical Learning: Data Mining, Inference And Prediction*, 2nd ed.; Springer: Berlin/Heidelberg, Germany, 2009.
33. Dietterich, T.G. Ensemble Methods in Machine Learning. In *International Workshop on Multiple Classifier Systems*; Springer: London, UK, 2000; pp. 1–15.
34. Maclin, R.; Opitz, D.W. Popular Ensemble Methods: An Empirical Study. *arXiv* **2011**, arXiv:1106.0257.
35. Breiman, L. Bagging Predictors. *Mach. Learn.* **1996**, *24*, 123–140.:1018054314350. [CrossRef]
36. Freund, Y.; Schapire, R.E. A Decision-Theoretic Generalization of On-Line Learning and an Application to Boosting. *J. Comput. Syst. Sci.* **1997**, *55*, 119–139. [CrossRef]
37. Datta, S.; Pihur, V.; Datta, S. An adaptive optimal ensemble classifier via bagging and rank aggregation with applications to high dimensional data. *BMC Bioinform.* **2010**, *11*, 427. [CrossRef] [PubMed]
38. Ahn, H.; Moon, H.; Fazzari, M.J.; Lim, N.; Chen, J.J.; Kodell, R.L. Classification by ensembles from random partitions of high-dimensional data. *Comput. Stat. Data Anal.* **2007**, *51*, 6166–6179. [CrossRef]
39. Lim, N.; Ahn, H.; Moon, H.; Chen, J.J. Classification of high-dimensional data with ensemble of logistic regression models. *J. Biopharm. Stat.* **2009**, *20*, 160–171. [CrossRef]
40. R Development Core Team. *R: A Language and Environment for Statistical Computing*; R Foundation for Statistical Computing: Vienna, Austria, 2008; ISBN 3-900051-07-0.
41. Kursa, M.B. *Praznik: Collection of Information-Based Feature Selection Filters*; R Package Version 5.0.0; R Foundation for Statistical Computing: Vienna, Austria, 2018.
42. Natalia Novoselova, Junxi Wang, F.P.F.K. *Biocomb: Feature Selection and Classification with the Embedded Validation Procedures for Biomedical Data Analysis*; R package version 0.4; R Foundation for Statistical Computing: Vienna, Austria, 2018.
43. Friedman, J.; Hastie, T.; Tibshirani, R. Regularization Paths for Generalized Linear Models via Coordinate Descent. *J. Stat. Softw.* **2010**, *33*, 1–22. [CrossRef]
44. Breheny, P.; Huang, J. Coordinate descent algorithms for nonconvex penalized regression, with applications to biological feature selection. *Ann. Appl. Stat.* **2011**, *5*, 232–253. [CrossRef]
45. Liaw, A.; Wiener, M. Classification and Regression by randomForest. *R News* **2002**, *2*, 18–22.
46. Meyer, D.; Dimitriadou, E.; Hornik, K.; Weingessel, A.; Leisch, F. *e1071: Misc Functions of the Department of Statistics, Probability Theory Group (Formerly: E1071), TU Wien*; R Package Version 1.7-1; R Foundation for Statistical Computing: Vienna, Austria, 2019.
47. Pihur, V.; Datta, S.; Datta, S. *RankAggreg: Weighted Rank Aggregation*; R package version 0.6.5; R Foundation for Statistical Computing: Vienna, Austria, 2018.
48. The RLFS-ERRM Resources 2019. Available online: https://sites.google.com/site/abhijeetrpatil01/file-cabinet/blfs-errm-manuscript-files-2019 (accessed on 25 December 2019).
49. Feature Selection Datasets. Available online: http://featureselection.asu.edu/old/datasets.php (accessed on 25 December 2019).
50. Bolón-Canedo, V.; Sánchez-Maroño, N.; Alonso-Betanzos, A.; Benítez, J.M.; Herrera, F. A review of microarray datasets and applied feature selection methods. *Inf. Sci.* **2014**, *282*, 111–135. [CrossRef]
51. Wang, M.; Barbu, A. Are screening methods useful in feature selection? An empirical study. *PloS ONE* **2018**, *14*, e0220842. [CrossRef]
52. Tsai, C.; Chen, J.J. Multivariate analysis of variance test for gene set analysis. *Bioinformatics* **2009**, *25*, 897–903. [CrossRef] [PubMed]
53. Li, J.; Cheng, K.; Wang, S.; Morstatter, F.; Trevino, R.P.; Tang, J.; Liu, H. Feature Selection: A Data Perspective. *ACM Comput. Surv.* **2017**, *50*, 94:1–94:45. [CrossRef]
54. Chen, X.D.; Ishwaran, H. Random forests for genomic data analysis. *Genomics* **2012**, *99*, 323–329. [CrossRef] [PubMed]
55. Bielza, C.; Robles, V.; Larrañaga, P. Regularized logistic regression without a penalty term: An application to cancer classification with microarray data. *Expert Syst. Appl.* **2011**, *38*, 5110–5118. [CrossRef]

56. Liao, J.G.; Chin, K.V. Logistic regression for disease classification using microarray data: model selection in a large p and small n case. *Bioinformatics* **2007**, *23*, 1945–1951. [CrossRef] [PubMed]
57. Zou, H.; Hastie, T. Regularization and Variable Selection via the Elastic Net. *J. R. Stat. Soc. Ser. B (Stat. Methodol.)* **2005**, *67*, 301–320. [CrossRef]
58. Li, Y.; Zhang, Y.; Zhao, S. Gender Classification with Support Vector Machines Based on Non-tensor Pre-wavelets. In Proceedings of the 2010 Second International Conference on Computer Research and Development, Kuala Lumpur, Malaysia, 7–10 May 2010; pp. 770–774.

© 2020 by the authors. Licensee MDPI, Basel, Switzerland. This article is an open access article distributed under the terms and conditions of the Creative Commons Attribution (CC BY) license (http://creativecommons.org/licenses/by/4.0/).

Article

Robust Linear Trend Test for Low-Coverage Next-Generation Sequence Data Controlling for Covariates

Jung Yeon Lee [1], Myeong-Kyu Kim [2] and Wonkuk Kim [3],*

[1] Department of Psychiatry, New York University School of Medicine, New York, NY 10016, USA; JungYeon.Lee@nyulangone.org
[2] Department of Neurology, Chonnam National University Medical School, Gwangju 61469, Korea; mkkim@jnu.ac.kr
[3] Department of Applied Statistics, Chung-Ang University, Seoul 06974, Korea
* Correspondence: wkim@cau.ac.kr; Tel.: +82-2-820-6688

Received: 30 December 2019; Accepted: 5 February 2020; Published: 8 February 2020

Abstract: Low-coverage next-generation sequencing experiments assisted by statistical methods are popular in a genetic association study. Next-generation sequencing experiments produce genotype data that include allele read counts and read depths. For low sequencing depths, the genotypes tend to be highly uncertain; therefore, the uncertain genotypes are usually removed or imputed before performing a statistical analysis. It may result in the inflated type I error rate and in a loss of statistical power. In this paper, we propose a mixture-based penalized score association test adjusting for non-genetic covariates. The proposed score test statistic is based on a sandwich variance estimator so that it is robust under the model misspecification between the covariates and the latent genotypes. The proposed method takes advantage of not requiring either external imputation or elimination of uncertain genotypes. The results of our simulation study show that the type I error rates are well controlled and the proposed association test have reasonable statistical power. As an illustration, we apply our statistic to pharmacogenomics data for drug responsiveness among 400 epilepsy patients.

Keywords: allele read counts; low-coverage; mixture model; next-generation sequencing; sandwich variance estimator

1. Introduction

Genome-wide association study (GWAS) is a powerful tool for screening a high-dimensional genome data set and selecting candidate genetic variants such as single nucleotide polymorphisms (SNPs) in genetic association studies. Next-generation sequencing (NGS) technology is widely used to produce a large amount of genetic information in a fast way. In the past decade, there have been numerous studies using NGS data such as rare variants association study [1,2], pharmacogenomics [3,4], machine learning and deep learning applications [5,6], and big data analysis [7,8]. Many NGS experiments are based on low-coverage sequencing with a large sized sample since there is a trade-off between sample size and sequencing depth in the NGS experiments [9,10]. For the low-coverage NGS data, a high uncertainty of the inferred genotypes is common; however, it causes biased and unreliable results on genetic association analyses. In genetic research based on NGS data, therefore, it is important to obtain accurate genotypes to perform an association analysis.

A number of researchers have worked on the effects of genotype misclassification in genetic association studies. There are two types of genotype misclassifications: differential and non-differential misclassifications, determined by whether the misclassification mechanism differs in the case and control groups or not. In summary, non-differential misclassifications result in a loss of statistical power and differential misclassifications distort type I error rates in a genetic case-control association study [11–14].

While there have been many research on improving the accuracy of genotypes such as the joint genotype calling algorithms across all samples were suggested to increase the accuracy of genotype calls [15–17], several researchers have tried to develop new association statistics accounting for the genotype errors. Their approaches are based on the raw measurements rather than inferred genotypes. In statistical genetics literature, Kim et al. [18] extended a chi-squared test of independence and developed a mixture likelihood based association test using the continuous measurements for copy number polymorphisms. Barnes et al. [19] proposed a mixture model linear trend test for the continuous copy number measurements. In NGS experiments, a likelihood ratio test based on allele read counts of pooled samples was proposed to test independence of genetic variants with a binary phenotype [20]. Gordon et al. [21] proposed a likelihood ratio test of the binomial mixture model of allele read counts with known error parameters. Kim et al. [13,22] proposed an extended version of Cochran–Armitage (CA) trend test and a multi-variant linear trend test for next-generation sequences data by using binomial mixture models. For a case-parent trio design, the binomial mixture model was applied to develop extended transmission disequilibrium tests (TDTs) based on read counts and read depths and to provide power analysis and sample size formulas [23]. All these approaches do not require genotype calls that can be highly uncertain when the read depth or coverage is low. However, none of these previous research has addressed how to include covariates in their mixture-based association studies.

When the covariates are independent of the latent genotypes, the extension of the mixture model based association tests is straightforward. However, if the latent genotype variable is associated with other covariates, then a likelihood based approach requires a model specification between the genotype variable and the other covariates as opposed to the previous research [16–23]. To our knowledge, this is the first study that investigates a genetic case-control association test controlling for covariates in low-coverage NGS experiments. Since we do not know the true model, we apply a sandwich variance estimator to develop a robust genetic association test statistic.

2. Materials and Methods

2.1. Mixture Model Accounting for Covariates

Let $\mathbf{w}$ be a covariate vector. Let y be a random variable indicating the case-control status of an individual such that $y = 1$ if a subject is in the case group and $y = 0$, otherwise. Let $\mathbf{z} = (z_0, z_1, z_2)$ denote an unobservable latent genotype vector, where $\sum_{g=0}^{2} z_g = 1$ and $z_g = 1$ if and only if the genotype is equal to g. Let x and v denote the minor allele read count and the read depth, respectively. The probability function is given by

$$
\begin{aligned}
p(y, x, v, \mathbf{w}) &= \sum_{\mathbf{z}} p(y, x, v, \mathbf{w}, \mathbf{z}) \\
&= \sum_{\mathbf{z}} p(x|v, \mathbf{w}, \mathbf{z}, y) p(y|v, \mathbf{w}, \mathbf{z}) p(\mathbf{z}|\mathbf{w}, v) p(\mathbf{w}, v) \\
&= p(\mathbf{w}, v) \sum_{\mathbf{z}} p(x|v, \mathbf{z}, y) p(y|\mathbf{z}, \mathbf{w}) p(\mathbf{z}|\mathbf{w}).
\end{aligned}
\tag{1}
$$

If the probability function of the read count x does not depend on the phenotype y, that is, $p(x|v, \mathbf{z}, y) = p(x|v, \mathbf{z})$, then it is called a non-differential error model. We apply a binary logit model to the case-control phenotype response variable y that is the same model for Cochran–Armitage trend test when perfect genotypes are available:

$$
p(y|\mathbf{z}, \mathbf{w}) = \frac{e^{y(\beta_s^T \mathbf{z} + \beta_w^T \mathbf{w})}}{1 + e^{\beta_0 + \beta_s^T \mathbf{z} + \beta_w^T \mathbf{w}}}.
\tag{2}
$$

We assume a binomial error model to the allele read counts as in previous research [13,16,20,21,23]:

$$p(x|v, \mathbf{z}, y) = \binom{v}{x} \left(\mathbf{u}_\epsilon^T \mathbf{z}\right)^x \left(1 - \mathbf{u}_\epsilon^T \mathbf{z}\right)^{v-x}, \quad (3)$$

where $\mathbf{u}_\epsilon = (\epsilon, 1/2, 1-\epsilon)^T$. For a differential error model, we can use $\mathbf{u}_\epsilon = y(\epsilon_1, 1/2, 1-\epsilon_1)^T + (1-y)(\epsilon_0, 1/2, 1-\epsilon_0)^T$. When perfect genotypes are available, we do not need the conditional probability of the genotype $\mathbf{z}$ given covariates $\mathbf{w}$ to perform genetic association tests since the logistic regression model is a conditional model given the genotypes and covariates. In this work, we assume a multinomial logit model for the latent genotype given the covariates as follows:

$$p(\mathbf{z}|\mathbf{w}) = \frac{\sum_{g=0}^{2} z_g e^{\gamma_g^T \mathbf{w}}}{\sum_{m=0}^{2} e^{\gamma_m^T \mathbf{w}}}, \quad (4)$$

where $\gamma_0 = (0,0,0)^T$ to remove over-parametrization. Other statistical models without the assumptions of a multinomial logit model may also be used for the relationship between covariates and latent genotypes, where we do not know the true model.

The likelihood function L and the log-likelihood function ℓ are written as

$$L = \prod_{k=1}^{N} \left[\sum_{\mathbf{z}_k} p(y_k|\mathbf{z}_k, \mathbf{w}_k) p(x_k|v_k, \mathbf{z}_k, y_k) p(\mathbf{z}_k|\mathbf{w}_k) p(\mathbf{w}_k, v_k) \right]$$

$$= \prod_{k=1}^{N} \sum_{i=0}^{2} \left\{ \left(\frac{e^{y_k(\beta s_i + \beta_w^T \mathbf{w}_k)}}{1 + e^{\beta s_i + \beta_w^T \mathbf{w}_k}} \right) \left(\binom{v_k}{x_k} (u_{\epsilon i})^{x_k} (1 - u_{\epsilon i})^{v_k - x_k} \right) \left(\frac{e^{\gamma_i^T \mathbf{w}_k}}{\sum_{m=0}^{2} e^{\gamma_m^T \mathbf{w}_k}} \right) p(\mathbf{w}_k, v_k) \right\}, \quad (5)$$

$$\ell = \sum_{k=1}^{N} \log \left[\sum_{i=0}^{2} \left\{ \left(\frac{e^{y_k(\beta s_i + \beta_w^T \mathbf{w}_k)}}{1 + e^{\beta s_i + \beta_w^T \mathbf{w}_k}} \right) \left((u_{\epsilon i})^{x_k} (1 - u_{\epsilon i})^{v_k - x_k} \right) \left(\frac{e^{\gamma_i^T \mathbf{w}_k}}{\sum_{m=0}^{2} e^{\gamma_m^T \mathbf{w}_k}} \right) \right\} \right]$$

$$+ \sum_{k=1}^{N} \log \binom{v_k}{x_k} p(\mathbf{w}_k, v_k). \quad (6)$$

The error parameter ϵ is commonly small and hence the estimate of ϵ is often equal to zero. The zero estimate of the error parameter results in a divergent information matrix. It prevents us from calculating Rao's score test statistic. In order to overcome this issue, we include a beta density penalty term to prevent from zero estimate of the error parameter. The penalized log-likelihood function is given by

$$\ell_p = \ell + C \log \left[\epsilon^{a_\epsilon} (1-\epsilon)^{b_\epsilon} \right]. \quad (7)$$

During this work, we choose $C = 1$ as in [24,25]. The penalized complete-data likelihood function is given by

$$L_C = \prod_{k=1}^{N} \prod_{i=0}^{2} \left[\frac{e^{y_k(\beta s_i + \beta_w^T \mathbf{w}_k)}}{1 + e^{\beta s_i + \beta_w^T \mathbf{w}_k}} \times \binom{v_k}{x_k} (u_{\epsilon i})^{x_k} (1 - u_{\epsilon i})^{v_k - x_k} \epsilon^{\frac{a_\epsilon}{N}} (1-\epsilon)^{\frac{b_\epsilon}{N}} \times \frac{e^{\gamma_i^T \mathbf{w}_k}}{\sum_{m=0}^{2} e^{\gamma_m^T \mathbf{w}_k}} \right]^{z_{ik}} \quad (8)$$

The complete data log-likelihood function is written as

$$\ell_C = \sum_{k=1}^{N} \sum_{i=0}^{2} z_{ik} \left[y_k(\beta s_i + \beta_w^T \mathbf{w}_k) - \log\left(1 + e^{\beta s_i + \beta_w^T \mathbf{w}_k}\right) \right]$$

$$+ \sum_{k=1}^{N} \sum_{i=0}^{2} z_{ik} \left[x_k \log(u_{\epsilon i}) + (v_k - x_k) \log(1 - u_{\epsilon i}) \right] + a_\epsilon \log \epsilon + b_\epsilon \log(1-\epsilon) \quad (9)$$

$$+ \sum_{k=1}^{N} \sum_{i=0}^{2} z_{ik} \left[\gamma_i^T \mathbf{w}_k - \log\left(\sum_{m=0}^{2} e^{\gamma_m^T \mathbf{w}_k} \right) \right].$$

2.2. Derivation of EM Algorithm under H_0

We apply the Expectation–Maximization (EM) algorithm [26] to estimate the parameters in our mixture model. Given data and the (r)-th step estimated parameters, the $(r+1)$-th E-step of the EM algorithm is written as

$$Q^{(r+1)} = \sum_{k=1}^{N}\sum_{i=0}^{2} \tau_{ik}^{(r)} \left[y_k(\beta s_i + \beta_w^T \mathbf{w}_k) - \log\left(1 + e^{\beta s_i + \beta_w^T \mathbf{w}_k}\right) \right]$$

$$+ \sum_{k=1}^{N}\sum_{i=0}^{2} \tau_{ik}^{(r)} \left[x_k \log(u_{\epsilon i}) + (v_k - x_k)\log(1 - u_{\epsilon i}) + a_\epsilon \log \epsilon + b_\epsilon \log(1 - \epsilon) \right] \quad (10)$$

$$+ \sum_{k=1}^{N}\sum_{i=0}^{2} \tau_{ik}^{(r)} \left[\gamma_i^T \mathbf{w}_k - \log\left(\sum_{m=0}^{2} e^{\gamma_m^T \mathbf{w}_k}\right) \right],$$

where

$$\tau_{ik}^{(r)} = \frac{\left(\frac{e^{y_k(\beta^{(r)} s_i + \beta_w^{(r)T} \mathbf{w}_k)}}{1+e^{\beta^{(r)} s_i + \beta_w^{(r)T} \mathbf{w}_k}}\right) \left((u_{\epsilon i}^{(r)})^{x_k}(1-u_{\epsilon i}^{(r)})^{v_k-x_k}\right) \left(\frac{e^{\gamma_i^{(r)T} \mathbf{w}_k}}{\sum_{m=0}^{2} e^{\gamma_m^{(r)T} \mathbf{w}_k}}\right)}{\sum_{g=0}^{2}\left[\left(\frac{e^{y_k(\beta^{(r)} s_g + \beta_w^{(r)T} \mathbf{w}_k)}}{1+e^{\beta^{(r)} s_g + \beta_w^{(r)T} \mathbf{w}_k}}\right)\left((u_{\epsilon g}^{(r)})^{x_k}(1-u_{\epsilon g}^{(r)})^{v_k-x_k}\right)\left(\frac{e^{\gamma_g^{(r)T} \mathbf{w}_k}}{\sum_{m=0}^{2} e^{\gamma_m^{(r)T} \mathbf{w}_k}}\right)\right]}. \quad (11)$$

We note that the posterior probability of subject k belonging to genotype class i depends on the all parameters. In M-step, the $(r+1)$-th estimates of the parameters are obtained by maximizing $Q^{(r+1)}$:

$$\frac{\partial Q^{(r+1)}}{\partial \beta} = \sum_{k=1}^{N}\sum_{i=0}^{2} \tau_{ik}^{(r)} s_i (y_k - \pi_{ik}) = 0 \quad (12)$$

$$\frac{\partial Q^{(r+1)}}{\partial \beta_w} = \sum_{k=1}^{N}\sum_{i=0}^{2} \tau_{ik}^{(r)} \mathbf{w}_k (y_k - \pi_{ik}) = 0 \quad (13)$$

$$\frac{\partial Q^{(r+1)}}{\partial \epsilon} = \sum_{k=1}^{N}\left[\tau_{0k}^{(r)}\left(\frac{x_k}{\epsilon} - \frac{v_k - x_k}{1-\epsilon}\right) + \tau_{2k}^{(r)}\left(\frac{v_k - x_k}{\epsilon} - \frac{x_k}{1-\epsilon}\right)\right] + \frac{a_\epsilon}{\epsilon} - \frac{b_\epsilon}{1-\epsilon} = 0 \quad (14)$$

$$\frac{\partial Q^{(r+1)}}{\partial \gamma_i} = \sum_{k=1}^{N} \mathbf{w}_k \left(\tau_{ik}^{(r)} - p_{ik}\right) = 0, \quad (15)$$

where we use notations $\pi_{ik} = \pi_{ik}(\beta, \beta_w) = \frac{e^{\beta s_i + \beta_w^T \mathbf{w}_k}}{1+e^{\beta s_i + \beta_w^T \mathbf{w}_k}}$ and $p_{ik} = p_{ik}(\gamma_1, \gamma_2) = \frac{e^{\gamma_i^T \mathbf{w}_k}}{\sum_{m=0}^{2} e^{\gamma_m^T \mathbf{w}_k}}$ for simplicity. From Equation (14), we derive a closed form iteration formula to update the allele read error parameter ϵ:

$$\epsilon^{(r+1)} = \frac{\sum_{k=1}^{N}\left[\tau_{0k}^{(r)} x_k + \tau_{2k}^{(r)}(v_k - x_k)\right] + a_\epsilon}{\sum_{k=1}^{N}\left[(\tau_{0k}^{(r)} + \tau_{2k}^{(r)}) v_k\right] + a_\epsilon + b_\epsilon}. \quad (16)$$

There is no closed form iteration formulas to update other parameters β, β_w, γ_i. The M-step for β, β_w, and γ can be obtained by the Newton–Raphson method. The Hessian matrix of $Q^{(r+1)}$ is given by

$$\frac{\partial^2 Q^{(r+1)}}{\partial \beta^2} = -\sum_{k=1}^{N}\sum_{i=0}^{2} \tau_{ik}^{(r)} s_i^2 \left[\pi_{ik}(1 - \pi_{ik})\right] \quad (17)$$

$$\frac{\partial^2 Q^{(r+1)}}{\partial \beta \partial \beta_w} = -\sum_{k=1}^{N}\sum_{i=0}^{2} \tau_{ik}^{(r)} s_i \mathbf{w}_k \left[\pi_{ik}(1 - \pi_{ik})\right] \quad (18)$$

$$\frac{\partial^2 Q^{(r+1)}}{\partial \beta_w \partial \beta_w^T} = -\sum_{k=1}^{N}\sum_{i=0}^{2} \tau_{ik}^{(r)} \mathbf{w}_k \mathbf{w}_k^T [\pi_{ik}(1-\pi_{ik})] \qquad (19)$$

$$\frac{\partial^2 Q^{(r+1)}}{\partial \gamma_i \partial \gamma_i^T} = -\sum_{k=1}^{N} \mathbf{w}_k \mathbf{w}_k^T [p_{ik}(1-p_{ik})] \qquad (20)$$

$$\frac{\partial^2 Q^{(r+1)}}{\partial \gamma_i \partial \gamma_j^T} = \sum_{k=1}^{N} \mathbf{w}_k \mathbf{w}_k^T [p_{ik} p_{jk}] \qquad (21)$$

$$\frac{\partial^2 Q^{(r+1)}}{\partial \gamma_i \partial \beta_w^T} = \frac{\partial^2 Q^{(r+1)}}{\partial \gamma_i \partial \beta} = 0 \qquad (22)$$

Let $M = \mathrm{diag}\left(\sum_{i=0}^{2} \tau_{ik}\pi_{ik}(1-\pi_{ik})\right)$ be an $N \times N$ diagonal matrix. Let $W = (w_{ik})$ be the $N \times p$ matrix of covariates. Let μ be an $N \times 1$ vector of $\mu_k = \sum_{ik} \tau_{ik}\pi_{ik}$ and Y be an $N \times 1$ vector of y_k. Initially, we set $\beta^{[0]} = \beta^{(r)}$ and update the parameter estimate by

$$\beta^{[t+1]} = \beta^{[t]} + (W^T M W)^{-1} W^T (Y - \mu). \qquad (23)$$

Let $D_{11} = \mathrm{diag}(p_{1k}(1-p_{1k}))$, $D_{12} = D_{21} = -\mathrm{diag}(p_{1k}p_{2k})$, and $D_{22} = \mathrm{diag}(p_{2k}(1-p_{2k}))$. Let $\tau_i = (\tau_{ik})$ be the $N \times 1$ vector and $p_i = (p_{ik})$ be the $N \times 1$ vector. Initially, set $\gamma_i^{[0]} = \gamma_i^{(r)}$ and update the parameters γ_i by

$$\begin{pmatrix} \gamma_1^{[t+1]} \\ \gamma_2^{[t+1]} \end{pmatrix} = \begin{pmatrix} \gamma_1^{[t]} \\ \gamma_2^{[t]} \end{pmatrix} + \begin{pmatrix} W^T D_{11} W & W^T D_{12} W \\ W^T D_{21} W & W^T D_{22} W \end{pmatrix}^{-1} \begin{pmatrix} W^T(\tau_1 - p_1) \\ W^T(\tau_2 - p_2) \end{pmatrix}. \qquad (24)$$

In order to obtain $\beta_w^{(r+1)}$ and $\gamma_i^{(r+1)}$, we stop the iterations in the M-step for β and γ_i when $||\beta^{[t+1]} - \beta^{[t]}||^2 + ||\gamma_1^{[t+1]} - \gamma_1^{[t]}||^2 + ||\gamma_2^{[t+1]} - \gamma_2^{[t]}||^2 \leq tol^2$ or the number of iterations reaches the prespecified maximum number of iterations. In our work, we set $tol = 10^{-6}$ and fix the maximum iteration as 1000.

2.3. Hypothesis Tests of Genetic Association Controlling for Covariates

To test genetic association between the latent genetic variables and the binary response variable while controlling covariates, we employ Rao's score test. There are several advantages for the use of the score test. Cochran-Armitage trend test with perfect genotypes is a score test, and we extend this test to when the genotypes are highly uncertain. The score test requires less computational cost compared to the likelihood ratio test since it requires the parameter estimates only under the null hypothesis of no association. The score function calculated in previous section is given by

$$S = \sum_{k=1}^{N}\sum_{i=0}^{2} \tau_{ik(0)} s_i \left(y_k - \frac{e^{\beta_{w(0)}^T \mathbf{w}_k}}{1 + e^{\beta_{w(0)}^T \mathbf{w}_k}} \right) \qquad (25)$$

where the subscript (0) denotes the estimated parameter under the null hypothesis. Another important issue to be considered when we include the covariates in a low-coverage next-generation sequencing genetic association study is a model misspecification of the latent genotypes on the covariates. To overcome this model misspecification problem, we employ the sandwich variance estimator [27]. In this work, we derive a robust generalized score test using the sandwich variance–covariance estimator. In general, one of the difficulties in applying the sandwich estimator in practice is that it requires analytic derivation for the covariance matrix of the proposed model. For simplicity in our derivation of the sandwich variance estimator, θ denotes the vector of all parameters $\theta = (\beta, \beta_w, \gamma, \epsilon)$, and $\phi = (\beta_w, \gamma, \epsilon)$ denotes the parameter vector except β, and hence $\theta = (\beta, \phi)$. The sandwich variance estimator for the score function S under H_0 is given by

$$v_s = V_{\beta\beta} - V_{\beta\phi} J_{\phi\phi}^{-1} J_{\phi\beta} - J_{\beta\phi} J_{\phi\phi}^{-1} V_{\phi\beta} + J_{\beta\phi} J_{\phi\phi}^{-1} V_{\phi\phi} J_{\phi\phi}^{-1} J_{\phi\beta}, \tag{26}$$

where $V = E_{f_0}\left[\frac{\partial \ell}{\partial \theta}\frac{\partial \ell}{\partial \theta}^T\right]$ and $J = -E_{f_0}\left[\frac{\partial^2 \ell}{\partial \theta \partial \theta^T}\right]$ under the unknown true distribution f_0. For simplicity, we may use h_{ik} during derivation of the sandwich variance estimator:

$$h_{gk} = \left(\frac{e^{y_k(\beta s_g + \beta_w^T w_k)}}{1 + e^{\beta s_g + \beta_w^T w_k}}\right)\left((u_{eg})^{x_k}(1 - u_{eg})^{v_k - x_k}\right)\left(\frac{e^{\gamma_g^T w_k}}{\sum_{m=0}^{2} e^{\gamma_m^T w_k}}\right), \tag{27}$$

so that the likelihood function is written as

$$\ell = \sum_{k=1}^{N} \log\left[\sum_{g=0}^{2} h_{gk}\right] + C. \tag{28}$$

The relationship between J and V can be written as

$$\begin{aligned}
J &= \frac{1}{N}\sum_{k=1}^{N}\left[\sum_{g=0}^{2} \tau_{gk}\frac{\partial}{\partial \theta}\log h_{gk}\right]\left[\sum_{g=0}^{2} \tau_{gk}\frac{\partial}{\partial \theta^T}\log h_{gk}\right] \\
&\quad - \frac{1}{N}\sum_{k=1}^{N}\sum_{g=0}^{2} \tau_{gk}\left[\left(\frac{\partial}{\partial \theta}\log h_{gk}\right)\left(\frac{\partial}{\partial \theta^T}\log h_{gk}\right) + \frac{\partial^2}{\partial \theta \partial \theta^T}\log h_{gk}\right] \\
&= V - \frac{1}{N}\sum_{k=1}^{N}\sum_{g=0}^{2} \tau_{gk}\left[\left(\frac{\partial}{\partial \theta}\log h_{gk}\right)\left(\frac{\partial}{\partial \theta^T}\log h_{gk}\right) + \frac{\partial^2}{\partial \theta \partial \theta^T}\log h_{gk}\right]
\end{aligned} \tag{29}$$

If there is no model misspecification, we have $J = V$ and the robust score test statistic is reduced to Rao's score test statistic. We denote the difference $R = V - J$ so that $R = \frac{1}{N}\sum_{k=1}^{N}\sum_{g=0}^{2} \tau_{gk}\left[\left(\frac{\partial}{\partial \theta}\log h_{gk}\right)\left(\frac{\partial}{\partial \theta^T}\log h_{gk}\right) + \frac{\partial^2}{\partial \theta \partial \theta^T}\log h_{gk}\right]$. The components of $\frac{\partial}{\partial \theta}\log h_{gk}$ are calculated by

$$\frac{\partial}{\partial \beta}\log h_{gk} = s_g[y_k - \pi_k] \tag{30}$$

$$\frac{\partial}{\partial \beta_w}\log h_{gk} = w_k[y_k - \pi_k] \tag{31}$$

$$\frac{\partial}{\partial \epsilon}\log h_{gk} = \delta_g(0)\left[\frac{X_k}{\epsilon} - \frac{V_k - X_k}{1 - \epsilon}\right] + \delta_g(2)\left[\frac{V_k - X_k}{\epsilon} - \frac{X_k}{1 - \epsilon}\right] + \frac{a_\epsilon}{N\epsilon} - \frac{b_\epsilon}{N(1 - \epsilon)} \tag{32}$$

$$\frac{\partial}{\partial \gamma_i}\log h_{gk} = w_k\left[I(g = i) - p_{ik}\right], \tag{33}$$

where $\delta_g(i) = 1$ if $g = i$ and $\delta_g(i) = 0$ if $g \neq i$. It is straightforward to calculate V from the above first derivatives. The second term $\frac{\partial^2}{\partial \theta \partial \theta^T}\log h_{gk}$ of R has components as

$$\frac{\partial^2}{\partial \beta^2}\log h_{gk} = -s_g^2 \pi_k(1 - \pi_k) \tag{34}$$

$$\frac{\partial^2}{\partial \beta_w \partial \beta_w^T}\log h_{gk} = -w_k w_k^T \pi_k(1 - \pi_k) \tag{35}$$

$$\frac{\partial^2}{\partial \beta_w \partial \beta}\log h_{gk} = -w_k s_g \pi_k(1 - \pi_k) \tag{36}$$

$$\frac{\partial^2}{\partial \gamma_i \partial \gamma_i^T}\log h_{gk} = -w_k w_k^T p_{ik}(1 - p_{ik}) \tag{37}$$

$$\frac{\partial^2}{\partial \gamma_i \partial \gamma_{3-i}^T} \log h_{gk} = \mathbf{w}_k \mathbf{w}_k^T p_{ik} p_{3-i,k} \tag{38}$$

$$\frac{\partial^2}{\partial \epsilon^2} \log h_{gk} = -\left(\delta_g(0) \left[\frac{X_k}{\epsilon^2} + \frac{V_k - X_k}{(1-\epsilon)^2}\right] + \delta_g(2) \left[\frac{V_k - X_k}{\epsilon^2} + \frac{X_k}{(1-\epsilon)^2}\right]\right.$$
$$\left. + \frac{a_\epsilon}{N\epsilon^2} + \frac{b_\epsilon}{N(1-\epsilon)^2}\right), \tag{39}$$

where $i = 1$ or 2. All other second derivatives that are not presented are equal to zero. Using these first and second derivatives of $\log h_{gk}$, we can obtain the components of the difference matrix R as follows:

$$R_{\beta\beta} = \frac{1}{N} \sum_{k=1}^{N} \sum_{g=0}^{2} \tau_{gk} s_g^2 \left[(y_k - \pi_k)^2 - \pi_k(1-\pi_k)\right] \tag{40}$$

$$R_{\beta w \beta} = \frac{1}{N} \sum_{k=1}^{N} \sum_{g=0}^{2} \tau_{gk} s_g \mathbf{w}_k \left[(y_k - \pi_k)^2 - \pi_k(1-\pi_k)\right] \tag{41}$$

$$R_{\beta w \beta w} = \frac{1}{N} \sum_{k=1}^{N} \mathbf{w}_k \mathbf{w}_k^T \left[(y_k - \pi_k)^2 - \pi_k(1-\pi_k)\right] \tag{42}$$

$$R_{\epsilon\epsilon} = \frac{1}{N} \sum_{k=1}^{N} \left(\tau_{0k} \left[\frac{X_k + a_\epsilon/N}{\epsilon} - \frac{V_k - X_k + b_\epsilon/N}{1-\epsilon}\right]^2 + \tau_{1k} \left[\frac{a_\epsilon}{N\epsilon} - \frac{b_\epsilon}{N(1-\epsilon)}\right]^2$$
$$+ \tau_{2k} \left[\frac{V_k - X_k + a_\epsilon/N}{\epsilon} - \frac{X_k + b_\epsilon/N}{1-\epsilon}\right]^2 - \tau_{0k} \left[\frac{X_k + a_\epsilon/N}{\epsilon^2} + \frac{V_k - X_k + b_\epsilon/N}{(1-\epsilon)^2}\right] \tag{43}$$
$$- \tau_{1k} \left[\frac{a_\epsilon}{N\epsilon^2} + \frac{b_\epsilon}{N(1-\epsilon)^2}\right] - \tau_{2k} \left[\frac{V_k - X_k + a_\epsilon/N}{\epsilon^2} + \frac{X_k + b_\epsilon/N}{(1-\epsilon)^2}\right]\right)$$

$$R_{\beta\epsilon} = \frac{1}{N} \sum_{k=1}^{N} [y_k - \pi_k] \left(\tau_{0k} s_0 \left[\frac{X_k + a_\epsilon/N}{\epsilon} - \frac{V_k - X_k + b_\epsilon/N}{1-\epsilon}\right] + \tau_{1k} s_1 \left[\frac{a_\epsilon}{N\epsilon} - \frac{b_\epsilon}{N(1-\epsilon)}\right]$$
$$+ \tau_{2k} s_2 \left[\frac{V_k - X_k + a_\epsilon/N}{\epsilon} - \frac{X_k + b_\epsilon/N}{1-\epsilon}\right]\right) \tag{44}$$

$$R_{\beta w\epsilon} = \frac{1}{N} \sum_{k=1}^{N} \mathbf{w}_k [y_k - \pi_k] \left(\tau_{0k} \left[\frac{X_k + \epsilon/N}{\epsilon} - \frac{V_k - X_k + b_\epsilon/N}{1-\epsilon}\right] + \tau_{1k} \left[\frac{a_\epsilon}{N\epsilon} - \frac{b_\epsilon}{N(1-\epsilon)}\right]$$
$$+ \tau_{2k} \left[\frac{V_k - X_k + a_\epsilon/N}{\epsilon} - \frac{X_k + b_\epsilon/N}{1-\epsilon}\right]\right) \tag{45}$$

$$R_{\gamma_i \gamma_i} = \frac{1}{N} \sum_{k=1}^{N} \mathbf{w}_k \mathbf{w}_k^T \left[(\tau_{ik} - p_{ik})(1 - 2p_{ik})\right] \tag{46}$$

$$R_{\gamma_1 \gamma_2} = \frac{1}{N} \sum_{k=1}^{N} \mathbf{w}_k \mathbf{w}_k^T \left[p_{1k}(p_{2k} - \tau_{2k}) + p_{2k}(p_{1k} - \tau_{1k})\right] \tag{47}$$

$$R_{\gamma_i \beta} = \frac{1}{N} \sum_{k=1}^{N} \mathbf{w}_k (y_k - \pi_k) \left[\tau_{ik} s_i - p_{ik} \sum_{g=0}^{2} \tau_{gk} s_g\right] \tag{48}$$

$$R_{\gamma_i \beta w} = \frac{1}{N} \sum_{k=1}^{N} \mathbf{w}_k \mathbf{w}_k^T (y_k - \pi_k) \left[\tau_{ik} - p_{ik}\right] \tag{49}$$

$$R_{\gamma_i \epsilon} = \frac{1}{N} \sum_{k=1}^{N} \mathbf{w}_k \left(-p_{1k} \tau_{0k} \left[\frac{X_k + a_\epsilon/N}{\epsilon} - \frac{V_k - X_k + b_\epsilon/N}{1-\epsilon}\right]\right.$$
$$+ (1 - p_{1k}) \tau_{1k} \left[\frac{a_\epsilon}{N\epsilon} - \frac{b_\epsilon}{N(1-\epsilon)}\right] \tag{50}$$
$$\left. - p_{1k} \tau_{2k} \left[\frac{V_k - X_k + a_\epsilon/N}{\epsilon} - \frac{X_k + b_\epsilon/N}{1-\epsilon}\right]\right)$$

$$R_{\gamma2\epsilon} = \frac{1}{N}\sum_{k=1}^{N} w_k \left(-p_{2k}\tau_{0k}\left[\frac{X_k + a_\epsilon/N}{\epsilon} - \frac{V_k - X_k + b_\epsilon/N}{1-\epsilon}\right] - p_{2k}\tau_{1k}\left[\frac{a_\epsilon}{N\epsilon} - \frac{b_\epsilon}{N(1-\epsilon)}\right]\right.$$
$$\left. + (1 - p_{2k})\tau_{2k}\left[\frac{V_k - X_k + \epsilon/N}{\epsilon} - \frac{X_k + b_\epsilon/N}{1-\epsilon}\right]\right) \tag{51}$$

Therefore, our proposed robust score test statistic Z_R can be written as

$$Z_R = \frac{S}{\sqrt{Nv_s}}, \tag{52}$$

which asymptotically has a standard normal distribution under H_0.

Another common approach to obtain p-values is to use Monte Carlo permutation method based on the score vector or function. However, the Monte Carlo permutation p-value calculation given a very small Bonferroni's corrected level of significance needs high computational expenses since it requires at least 10^7 or 10^8 permuted resamples. In this work, we employ the asymptotic permutation p-value calculation. The score function is given by

$$S = \sum_{k=1}^{N}\sum_{i=0}^{2} \tau_{ik(0)} s_i \left(y_k - \frac{e^{\beta_{w(0)}^T w_k}}{1 + e^{\beta_{w(0)}^T w_k}}\right)$$
$$= \sum_{k=1}^{N} r_k e_k \tag{53}$$

where the subscript (0) denotes the estimated parameter under the null hypothesis. We define a score $r_k = \sum_{i=0}^{2} \tau_{ik(0)} s_i$ associated with subject k and the kth residual $e_k = \left(y_k - \frac{e^{\beta_{w(0)}^T w_k}}{1 + e^{\beta_{w(0)}^T w_k}}\right)$. We can permute the residuals e_k's to calculate the permutation p-value for adjusting covariate effects. The asymptotic permutation test statistic Z_{AP} for a large sample size is given by

$$Z_{AP} = \frac{S - N \cdot \bar{r} \cdot \bar{e}}{\sqrt{\frac{1}{N-1}\left[\sum_{i=1}^{N} e_i^2 - N(\bar{e})^2\right]\left[\sum_{i=1}^{N} r_i^2 - N(\bar{r})^2\right]}} \tag{54}$$

where $\bar{r} = \frac{1}{N}\sum_{i=1}^{N} r_i$ and $\bar{e} = \frac{1}{N}\sum_{i=1}^{N} e_i$. The simple linear rank test statistic Z_{AP} asymptotically has a standard normal distribution under the null hypothesis [28].

3. Results

3.1. Simulation Study

In this section, we simulate data from the following process:

$$P(Y=1|w)f(w) = \sum_{i=0}^{2} P(Y=1|G=i,w)P(G=i)f(w) \tag{55}$$

For simplicity, we assume genetic relative risk $R_i = \frac{P(Y=1|G=i,w)}{P(Y=1|G=0,w)}$, for $i=1,2$, does not depend on the covariate W. We assume that the genotype frequency $\pi_i = P(G=i)$ satisfies Hardy–Weinberg equilibrium (HWE), so that $P(G=0) = p^2, P(G=1) = 2pq$, and $P(G=2) = q^2$, where q is the minor allele frequency. Then, the prevalence is given by

$$\phi = \int P(Y=1|w)f(w)dw$$
$$= \int \left[p^2 f(w|G=0) + 2pqR_1 f(w|G=1) + q^2 R_2 f(w|G=2)\right] P(Y=1|G=0,w)dw \tag{56}$$

We consider two scenarios when generating covariates w: (1) $f(w|G = i)$ is equal to a standard normal $N(0,1)$ for all $i = 0,1,2$, called by a single normal, and (2) $f(w|G = i)$ has a normal distribution with mean μ_i and standard deviation $\sigma = 1$, we call this a normal mixture. For the single normal model,

$$\phi = \left[p^2 + 2pqR_1 + q^2R_2\right] \int P(Y = 1|G = 0, w) f(w) dw \tag{57}$$

We finally assume $P(Y = 1|G = 0, w) = \frac{e^{\alpha+\beta_w w}}{1+e^{\alpha+\beta_w w}}$. During the simulation study, we compute α by numerical integration given prevalence ϕ and other parameters.

3.1.1. Simulation Study for Null Distribution

To evaluate the type I error rate of the proposed test statistic, we perform simulations with 5000 replicates per each parameter setting. We fixed the proportion of cases as 0.5. The parameter settings that we consider are:

(i) Prevalence (ϕ): 0.1, 0.3
(ii) Coverage (v): 4, 30
(iii) Minor allele frequency (q): 0.05, 0.3
(iv) Total sample size (n): 500, 1000, 1500
(v) Covariate (w_1): single normal or normal mixture with mean $\mu = (0, \frac{1}{2}, \frac{1}{2})$ given genotype $(0,1,2)$
(vi) Regression coefficient β_w: 0, 1

We consider prevalence $\phi = 0.3$ that may be large in a genetic association study. It is chosen to reflect pharmacogenomics data that we use in the real data analysis.

Figure 1 shows boxplots of the null simulations. The permutation method appears to have more variability of the empirical rejection rates over different configurations and to have the smaller empirical rejection rates compared to the proposed robust score test based on the sandwich variance estimator. When the sample size was small as 500 and the coverage was 4×, the permutation-based test had less than 2.5% rejection rate though the desired value is 5%. The smallest empirical rejection rate for the proposed robust test was greater than 3.5%, and it appears the empirical rejection rates become closer to 5% as the sample size increases. If the coverage is 30× or higher, then the estimated posterior probabilities in our approach are close to zero-or-one and most inferred genotypes are quite clear. When the coverage was 30×, our proposed test seems to well control the type I error rates regardless of other parameter settings as expected. Table 1 shows the empirical rejection rates under the null settings by combining our simulation results for the lower level of significance.

Table 1. Empirical rejection rates under null settings for level $1 \times 10^{-2}, 1 \times 10^{-3}, 1 \times 10^{-4}$, and 1×10^{-5}.

Method (cvrg)	1×10^{-2}	1×10^{-3}	1×10^{-4}	1×10^{-5}
Permutation (4×)	7.13×10^{-3}	6.14×10^{-4}	4.44×10^{-5}	0
Permutation (30×)	7.73×10^{-3}	7.08×10^{-4}	6.11×10^{-5}	5.56×10^{-6}
Sandwich (4×)	8.34×10^{-3}	6.89×10^{-4}	4.44×10^{-5}	5.56×10^{-6}
Sandwich (30×)	1.02×10^{-2}	1.01×10^{-3}	8.75×10^{-5}	8.33×10^{-6}

3.1.2. Simulation Study for Statistical Power

We used the same parameter settings as in the null simulation study. Additionally, we set multiplicative genetic relative risks vector $(1, 1.5, 1.5^2)$ in the alternative parameter configurations. In the alternative simulations, we calculated empirical rejection rates under Bonferroni corrected level of significance, that is, 5×10^{-8}. Figure 2 shows the boxplots of empirical power under various alternative settings. We removed the results when the sample size was 500 or the minor allele frequency was 0.05 since all the rejection rates were small in Figure 2. It appears interesting that the power of the proposed test when the coverage was 4× and the sample size was 1500 is higher than the power of the test when the coverage was 30× and the sample size was 1000. If the two design costs are similar, then the low-coverage with more samples seems more effective than the high-coverage with less samples.

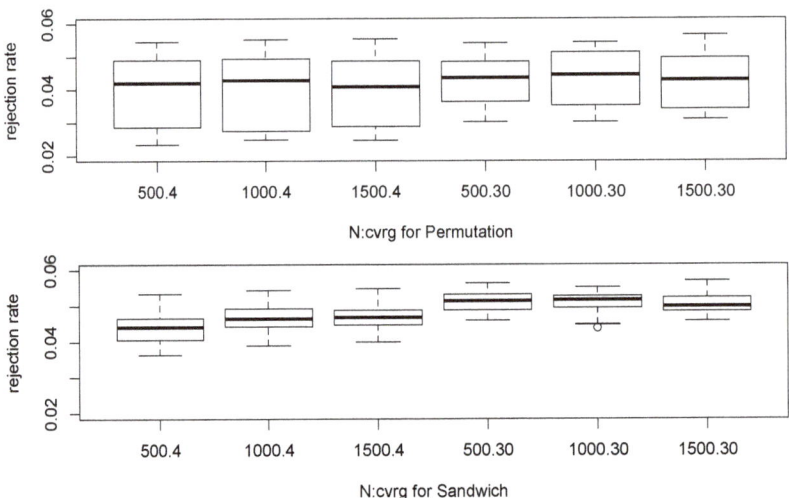

Figure 1. Boxplot of the empirical rejection rates under the null hypothesis.

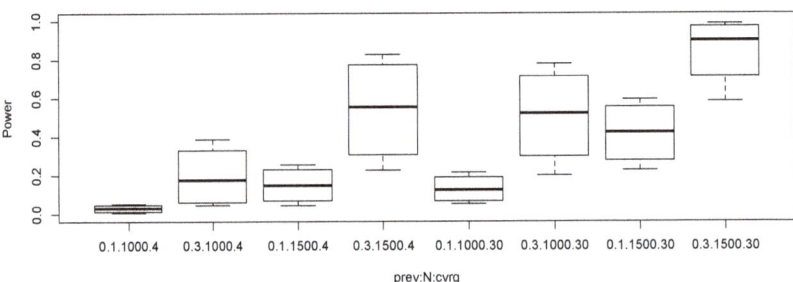

Figure 2. Boxplots of statistical power of the proposed robust test under the alternative settings. The level of significance was set as 5×10^{-8}. The notation 0.1.1000.4 represents prevalence 0.1, total sample size 1000, and coverage 4×.

Table 2 summarizes statistical power of our proposed method and a naive approach. The naive approach uses uncertain genotypes by the maximum posterior probability classification rule [29]. The standard logistic regression was applied to the uncertain genotypes. As expected, the proposed robust method shows higher power than the naive approach when the sequencing coverage is as low as 4×. When the sequencing coverage is high as 30×, two approaches show similar performance in terms of statistical power.

Table 2. Empirical rejection rates under alternative hypothesis. The level of significance was set as 5×10^{-8}.

Coverage	Total Sample Size	Covariate	β_w	Naive	Proposed
4	1000	Normal mixture	0	0.102	0.113
4	1000	Normal mixture	1	0.233	0.261
4	1000	Single normal	0	0.190	0.277
4	1000	Single normal	1	0.269	0.374
4	1500	Normal mixture	0	0.398	0.429
4	1500	Normal mixture	1	0.657	0.701
4	1500	Single normal	0	0.626	0.741
4	1500	Single normal	1	0.736	0.840
30	1000	Normal mixture	0	0.384	0.355
30	1000	Normal mixture	1	0.617	0.603
30	1000	Single normal	0	0.622	0.637
30	1000	Single normal	1	0.734	0.760
30	1500	Normal mixture	0	0.792	0.761
30	1500	Normal mixture	1	0.959	0.954
30	1500	Single normal	0	0.933	0.939
30	1500	Single normal	1	0.978	0.978

3.2. Real Data Analysis

The proposed robust generalized score test was applied to the pharmacogenomics data consisting of 400 epilepsy patients [22]. The data were collected from several epilepsy clinics in Korea and were genotyped for whole-exomes by NGS experiments [30]. All study participants followed the criteria in [31] if the participants had drug-resistant (case group) or drug-responsive (control group) epilepsy. We defined the drug resistance as the occurrence of at least four unprovoked seizures during the past one year at the time of recruitment, with trials of two or more appropriate antiepileptic drugs (AEDs) at maximal tolerated doses. Patients who underwent surgical treatment for drug-resistant epilepsy were classified as having drug-resistant epilepsy, regardless of the surgical outcome. We excluded some patients from the study if they were frequently in poor compliance with AED therapy and had reported seizures with a questionable semiology. In addition, we defined the drug responsiveness as complete freedom from seizures for at least one year up to the date of the last follow-up visit.

We included two non-genetic covariates in our association analysis. The two covariates were age of patient and duration of epileptic seizures. The age variable was definitely independent of genetic information, whereas duration variable may be associated with genetic variables. Due to the relatively small sample size 400, we did not expect to find a significantly associated SNP controlling for the two covariates. Therefore, instead of reporting a genome-wide association study, we illustrated the results of a SNP with low read depths and a SNP with high read depths. For the low read depths example, we selected a SNP from chromosome 1, which is rs3811406. The distribution of read depths for the SNP was summarized in Table 3. More than 10% of the sample had five or less read depths and more than 30% of the sample had 10 or less read depths at the SNP. When applying our proposed mixture-based association test, the test statistic value was $z_R = 2.864$ and the p-value was $p = 4.183 \times 10^{-3}$, while the standard logistic regression analysis using pooled genotype calls had $z = 2.601$ and the p-value $p = 9.30 \times 10^{-3}$ that was more than twice the p-value of the proposed robust test.

Table 3. Distribution of read depths at rs3811406.

Read Depth v	$v \leq 5$	$5 < v \leq 10$	$10 < v \leq 30$	$v > 30$	Total
Frequency	43	86	95	176	400
Proportion	0.1075	0.215	0.2375	0.44	1

In addition, we applied our proposed test to SNP *rs4915154* at which all patients had 13× or higher read depths and 85% patients had 25× or higher read depths. For this SNP, the proposed robust test statistic was $z_R = 2.940$ with p-value $= 3.28 \times 10^{-3}$ and the multiple logistic regression with the pooled genotype calls reported $z = 2.963$ with p-value $= 3.05 \times 10^{-3}$. The two results were quite close, as expected, due to high read depths at the SNP.

4. Discussion and Conclusions

In the present study, we developed the mixture-based genetic association tests adjusting the effects of non-genetic covariates in low-coverage NGS data. In order to construct a robust test statistic under model misspecification, we derived the sandwich variance estimator of the mixture model. The proposed test statistic is calculated from allele read counts and read depths instead of inferred genotypes so that we can apply this association test to low-coverage NGS data controlling for non-genetic covariates without external imputation or elimination of uncertain genotypes. Another important issue that we addressed in the present study is that the proposed test takes account of potential dependence between latent genotypes and the non-genetic covariates. Regarding computational cost, our proposed method is efficient because it is a generalized score test that uses the estimates of the parameters only under the null hypothesis of no association. When the sequencing depth is 4×, it takes around 1.2 s for sample size 500, 4 s for sample size 1000, and 9 s for sample size 1500 to simulate a dataset and to calculate both test statistics Z_{AP} and Z_R. When the sequencing depth is 30×, it takes approximately 0.13 s for sample size 500, 0.3 s for sample size 1000, and 0.53 s for sample size 1500. Time for these computations is measured based on a single core work of a 3.5 GHz Intel Xeon processor. As illustrated in the real data analysis section, the read depth is not a fixed constant. Therefore, the computational time for real data is usually less than that for the coverage 4× simulation setting. We used statistical software R, which is known to be slow. It would be computationally beneficial to run our proposed methods in other faster program languages for a high-dimensional genome-wide association study.

We applied the penalized likelihood method to avoid singularity of information matrix when calculating the proposed score test statistic. Therefore, the penalty term is not necessary for a non-zero estimate of the error parameter. During our work, we fixed the degree of penalization $C = 1$, $a_\epsilon = 0.01$, and $b_\epsilon = 0.99$ that implies 1% of allele read error as prior information. This parameter choice does not affect the proposed test statistic much since the likelihood function is merely changed when the sample size is greater than 500. It may be of interest to find optimal values for the parameters of the penalty term.

The simulation study confirms that the type I error rates of the proposed test are well controlled under the various parameter settings. The proposed robust test appears to perform better than the permutation based approach. Simulation results indicate that coverage 4× with sample size 1500 shows higher power as compared to coverage 30× with sample size 1000. Our method can be applied to an NGS experimental design by simulations to select coverage and sample size given a fixed amount of budget.

We presented a real data example in which the proposed test and multiple logistic regression results are similar to one another if the sequencing depth is high, whereas the test results may differ when the sequencing depth is low. This might have been caused because the proposed test is an extension of the multiple logistic regression with the unobserved latent genotype predictor. If the sequencing depth is high enough to call accurate genotypes, then our probability model becomes identical to the probability model of the multiple logistic regression. It would be more beneficial to compare with the previous methods by evaluating our proposed methods using a larger sized public dataset.

In this work, we focused on a single variant association test while controlling covariates. By adopting a multivariate mixture model, the proposed method can be extended to the multi-variant genetic association test including covariates. We can also extend the present method to differential genotype misclassifications.

Author Contributions: Conceptualization, W.K.; methodology, J.Y.L. and W.K.; software, J.Y.L. and W.K.; formal analysis, J.Y.L. and W.K.; data curation, M.-K.K. and W.K.; writing—original draft preparation, J.Y.L. and W.K.; writing—review and editing, J.Y.L., M.-K.K., and W.K.; project administration, W.K.; funding acquisition, M.-K.K. and W.K. All authors have read and agreed to the published version of the manuscript.

Funding: This research was supported by Basic Science Research Program through the National Research Foundation of Korea (NRF) funded by the Ministry of Education (NRF-2018R1D1A1B07050012) and was supported by a grant of the Korea Health Technology R&D Project through the Korea Health Industry Development Institute, funded by the Ministry of Health & Welfare, Republic of Korea (HI15C1559).

Conflicts of Interest: The authors declare no conflict of interest.

Abbreviations

The following abbreviations are used in this manuscript:

EM	Expectation–Maximization
GWAS	Genome-wide association study
HWE	Hardy–Weinberg equilibrium
maf	Minor allele frequency
NGS	Next-generation sequence
SNP	Single nucleotide polymorphism
TDT	Transmission disequilibrium test

References

1. Wu, M.C.; Lee, S.; Cai, T.; Li, Y.; Boehnke, M.; Lin, X. Rare-variant association testing for sequencing data with the sequence kernel association test. *Am. J. Hum. Genet.* **2011**, *89*, 82–93. [CrossRef]
2. Cirulli, E.T.; White, S.; Read, R.W.; Elhanan, G.; Metcalf, W.J.; Tanudjaja, F.; Fath, D.M.; Sandoval, E.; Isaksson, M.; Schlauch, K.A.; et al. Genome-wide rare variant analysis for thousands of phenotypes in over 70,000 exomes from two cohorts. *Nat. Commun.* **2020**, *11*, 542. [CrossRef]
3. Lakiotaki, K.; Kanterakis, A.; Kartsaki, E.; Katsila, T.; Patrinos, G.P.; Potamias, G. Exploring public genomics data for population pharmacogenomics. *PLoS ONE* **2017**, *12*, e0182138. [CrossRef] [PubMed]
4. Patrinos, G.P.; Giannopoulou, E.; Katsila, T.; Tsermpini, E.E.; Mitropoulou, C. Integrating next-generation sequencing in the clinical pharmacogenomics workflow. *Front. Pharmacol.* **2019**, *10*, 384.
5. Celesti, F.; Celesti, A.; Wan, J.; Villari, M. Why Deep Learning Is Changing the Way to Approach NGS Data Processing: A Review. *IEEE Rev. Biomed. Eng.* **2018**, *11*, 68–76. [CrossRef]
6. Le, N.Q.K.; Yapp, E.K.Y.; Nagasundaram, N.; Chua, M.C.H.; Yeh, H.Y. Computational identification of vesicular transport proteins from sequences using deep gated recurrent units architecture. *Comput. Struct. Biotechnol. J.* **2019**, *17*, 1245–1254. [CrossRef]
7. Tripathi, R.; Sharma, P.; Chakraborty, P.; Varadwaj, P.K. Next-generation sequencing revolution through big data analytics. *Front. Life Sci.* **2016**, *9*, 119–149. [CrossRef]
8. Cirillo, D.; Valencia, A. Big data analytics for personalized medicine. *Curr. Opin. Biotechnol.* **2019**, *58*, 161–167. [CrossRef]
9. Sims, D.; Sudbery, I.; Ilott, N.E.; Heger, A.; Ponting, C.P. Sequencing depth and coverage: Key considerations in genomic analyses. *Nat. Rev. Genet.* **2014**, *15*, 121–132. [CrossRef]
10. Song, K.; Li, L.; Zhang, G. Coverage recommendation for genotyping analysis of highly heterologous species using next-generation sequencing technology. *Sci. Rep.* **2016**, *6*, 35736. [CrossRef]
11. Gordon, D.; Finch, S.J.; Nothnagel, M.; Ott, J. Power and sample size calculations for case-control genetic association tests when errors are present: Application to single nucleotide polymorphisms. *Hum. Hered.* **2002**, *54*, 22–33. [CrossRef]
12. Ahn, K.; Haynes, C.; Kim, W.; Fleur, R.S.; Gordon, D.; Finch, S.J. The effects of SNP genotyping errors on the power of the Cochran-Armitage linear trend test for case/control association studies. *Ann. Hum. Genet.* **2007**, *71*, 249–261. [CrossRef]

13. Kim, W.; Londono, D.; Zhou, L.; Xing, J.; Nato, A.Q.; Musolf, A.; Matise, T.C.; Finch, S.J.; Gordon, D. Single-variant and multi-variant trend tests for genetic association with next-generation sequencing that are robust to sequencing error. *Hum. Hered.* **2012**, *74*, 172–183. [CrossRef]
14. Hou, L.; Sun, N.; Mane, S.; Sayward, F.; Rajeevan, N.; Cheung, K.; Cho, K.; Pyarajan, S.; Aslan, M.; Miller, P. Impact of genotyping errors on statistical power of association tests in genomic analyses: A case study. *Genet. Epidemiol.* **2017**, *41*, 152–162. [CrossRef]
15. Consortium, .G.P. An integrated map of genetic variation from 1092 human genomes. *Nature* **2012**, *491*, 56.
16. Le, S.Q.; Durbin, R. SNP detection and genotyping from low-coverage sequencing data on multiple diploid samples. *Genome Res.* **2011**, *21*, 952–960. [CrossRef]
17. Li, Y.; Sidore, C.; Kang, H.M.; Boehnke, M.; Abecasis, G.R. Low-coverage sequencing: Implications for design of complex trait association studies. *Genome Res.* **2011**, *21*, 940–951. [CrossRef]
18. Kim, W.; Gordon, D.; Sebat, J.; Kenny, Q.Y.; Finch, S.J. Computing power and sample size for case-control association studies with copy number polymorphism: Application of mixture-based likelihood ratio test. *PLoS ONE* **2008**, *3*, e3475. [CrossRef]
19. Barnes, C.; Plagnol, V.; Fitzgerald, T.; Redon, R.; Marchini, J.; Clayton, D.; Hurles, M.E. A robust statistical method for case-control association testing with copy number variation. *Nat. Genet.* **2008**, *40*, 1245. [CrossRef]
20. Kim, S.Y.; Li, Y.; Guo, Y.; Li, R.; Holmkvist, J.; Hansen, T.; Pedersen, O.; Wang, J.; Nielsen, R. Design of association studies with pooled or un-pooled next-generation sequencing data. *Genet. Epidemiol.* **2010**, *34*, 479–491. [CrossRef]
21. Gordon, D.; Finch, S.J.; De La Vega, F. A new expectation-maximization statistical test for case-control association studies considering rare variants obtained by high-throughput sequencing. *Hum. Hered.* **2011**, *71*, 113–125. [CrossRef]
22. Kim, W.; Kim, Y.H. Genetic association tests when a nuisance parameter is not identifiable under no association. *Commun. Stat. Appl. Methods* **2017**, *24*, 663–671. [CrossRef]
23. Kim, W. Transmission Disequilibrium Tests Based on Read Counts for Low-Coverage Next,-Generation Sequence Data. *Hum. Hered.* **2015**, *80*, 36–49. [CrossRef]
24. Chen, H.; Chen, J.; Kalbfleisch, J.D. A modified likelihood ratio test for homogeneity in finite mixture models. *J. R. Stat. Soc. Ser. B* **2001**, *63*, 19–29. [CrossRef]
25. Zhou, H.; Pan, W. Binomial mixture model-based association tests under genetic heterogeneity. *Ann. Hum. Genet.* **2009**, *73*, 614–630. [CrossRef]
26. Dempster, A.P.; Laird, N.M.; Rubin, D.B. Maximum likelihood from incomplete data via the EM algorithm. *J. R. Stat. Soc. Ser. B* **1977**, *39*, 1–22.
27. White, H. Maximum Likelihood Estimation of Misspecified Models. *Econometrica* **1982**, *50*, 1–25. [CrossRef]
28. Sidak, Z.; Sen, P.K.; Hajek, J. *Theory of Rank Tests*; Academic Press: San Diego, CA, USA, 1999.
29. Anderson, T.W. *An Introduction to Multivariate Statistical Analysis*; Wiley: New York, NY, USA, 1962.
30. Kang, K.W.; Kim, W.; Cho, Y.W.; Lee, S.K.; Jung, K.Y.; Shin, W.; Kim, D.W.; Kim, W.J.; Lee, H.W.; Kim, W. Genetic characteristics of non-familial epilepsy. *PeerJ* **2019**, *7*, e8278. [CrossRef]
31. Kim, M.-K.K.; Moore, J.H.; Kim, J.K.; Cho, K.H.; Cho, Y.W.; Kim, Y.S.; Lee, M.C.; Kim, Y.O.; Shin, M.H. Evidence for epistatic interactions in antiepileptic drug resistance. *J. Hum. Genet.* **2011**, *56*, 71–76. [CrossRef]

© 2020 by the authors. Licensee MDPI, Basel, Switzerland. This article is an open access article distributed under the terms and conditions of the Creative Commons Attribution (CC BY) license (http://creativecommons.org/licenses/by/4.0/).

Article

Comparing Groups of Decision-Making Units in Efficiency Based on Semiparametric Regression

Hohsuk Noh [1] and Seong J. Yang [2,*]

1. Department of Statistics, Sookmyung Women's University, Seoul 04310, Korea; hsnoh@sookmyung.ac.kr
2. Department of Statistics (Institute of Applied Statistics), Jeonbuk National University, Jeollabuk-do 54896, Korea
* Correspondence: sjyang@jbnu.ac.kr

Received: 26 December 2019; Accepted: 7 February 2020; Published: 11 February 2020

Abstract: We consider a stochastic frontier model in which a deviation of output from the production frontier consists of two components, a one-sided technical inefficiency and a two-sided random noise. In such a situation, we develop a semiparametric regression-based test and compare the technical efficiencies of the different decision-making unit groups, assuming that the production frontier function is the same for all the groups. Our test performs better than the previously proposed ones for the same purpose in numerical studies, and also has the theoretical advantage of working under more general assumptions. To illustrate our method, we apply the proposed test to Program for International Student Assessment (PISA) 2015 data and investigate whether an efficiency difference exists between male and female student groups at a specific age in terms of learning time and achievement in mathematics.

Keywords: data envelopment analysis; stochastic frontier model; semiparametric regression; group efficiency comparison

1. Introduction

Efficiency comparison between groups is currently used in various fields such as banking, insurance, sports, and R&D investment evaluation. Numerous empirical studies frequently analyze group efficiency using so-called Data Envelopment Analysis (DEA). DEA is a body of techniques for measuring relative efficiency by comparing it with the possible frontiers of decision-making units (DMUs) with multiple inputs and outputs. Here, the term DMU is used to collectively refer to all the units in which the production activity takes place. In the DEA framework, the DMU efficiency scores of each group can be obtained after specifying some assumptions appropriate to the situation, and then the comparison of the efficiency distributions of the groups is made on the basis of their obtained scores. For example, Golany and Storberg [1] and Lee et al. [2] applied non-parametric tests, such as the Mann–Whitney (MW) and Kruskal–Wallis tests, to the efficiency scores. Cummins et al. [3] introduced a dummy variable to indicate the groups, and then regressed the efficiency scores on the dummy variable. Simar and Zelenyuk [4] adapted the test developed in Li [5] to the DEA context and applied it to the obtained scores, to test the equality of efficiency distributions. O'Donnell et al. [6] used the concept of a meta-frontier to compare the technical efficiencies of firms that may be classified into different groups.

However, this stream of research under the DEA framework has a limitation in that it does not consider the noise factor in the production process. DEA typically assumes that the inefficiency of the DMU is the only cause of its production not reaching its maximum output, but obviously there are many uncontrolled factors which need to be considered as the cause. From this recognition, Aigner and Chu [7] and Meeusen and van den Broeck [8] first proposed the stochastic frontier model (SFM), which allows for both unobserved variation in output: the technical inefficiency of the production unit and the noise which represents the effect of innumerable uncontrollable factors. For illustrative comparison between DEA and SFM frameworks, see Figure 1.

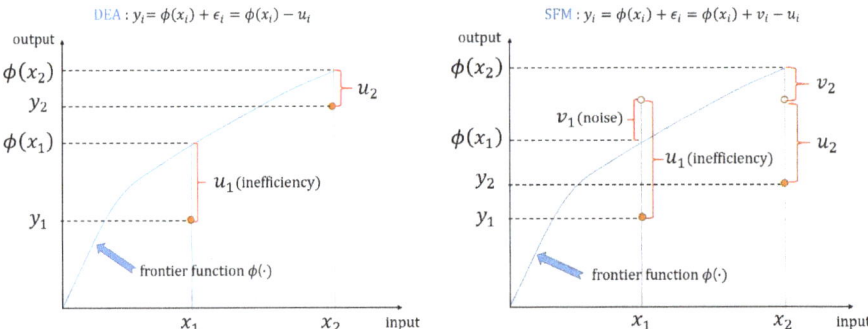

Figure 1. Comparison between Data Environment Analysis (DEA) (**left panel**) and SFM (**right panel**) frameworks (y_i: output, x_i: input, $\phi(x_i)$: the maximum output which can be obtained from the input x_i, ϵ_i: deviation from the production frontier function $\phi(x_i)$, u_i: technical inefficiency, v_i: noise). Note that the technical inefficiency u_i is a nonnegative random variable with unknown distribution.

Nowadays, the stochastic frontier model is used in a large literature of studies of production. Hence, we feel the need to develop a method and compare the efficiency difference between groups under SFM framework. One pioneering work in this direction is Banker et al. [9]. They developed five DEA-based hypothesis tests to compare the efficiency of groups under SFM. Although the paper referred above is an important development toward group efficiency comparison under SFM, their tests need to improve further.

First, their rather strong assumptions might limit the applicability of the proposed methods. For their parametric tests, they assumed the equality of both noise variance and inefficiency variance across groups. Second, their theoretical justification of the proposed methods needs to be checked. As regards their ordinary least squares (OLS) test of the mean difference in inefficiency, they provided its asymptotic normality as theoretical basis, but to our knowledge, such asymptotic normality is difficult to obtain because of the slow convergence rate of the DEA estimator when the number of input variables is greater than or equal to 2. The same comment is given in Section 3.2 of Simar and Wilson [10] on a similar type of asymptotic normality result as proof of Proposition 1 in Banker and Natarajan [11]. Finally, because they used the DEA methods for SFM, the tests they developed were based not directly on inefficiency itself, but on the inefficiency contaminated by positive measurement error due to noise. This indirect approach can lower the performance of their tests.

This observation has motivated us to develop a theoretically sound tool for comparison of group inefficiencies in the presence of noise. We develop such a methodology using a semi-parametric regression technique instead of DEA methods. The newly developed test performs better than the tests of Banker et al. [9] in numerical studies. It also has the theoretical advantage of working under more general assumptions compared to Banker et al. [9].

The rest of this paper is organized as follows. Section 2 describes our proposed test for group inefficiency comparison. We then perform some simulation studies and compare our test with the tests proposed by Banker et al. [9] in Section 3. We illustrate our method by applying the proposed test to Program for International Student Assessment (PISA) 2015 data and investigate whether an efficiency difference exists between male and female student groups at a specific age in terms of learning time and achievement in mathematics in Section 4. Section 5 provides some discussion and future research topics.

2. Group Efficiency Comparison under SFM

Assume that we have observations on n DMUs, where each observation consists of a vector of p inputs $\mathbf{X}_i = (X_{1,i}, \ldots, X_{p,i})^\top$ and the corresponding output Y_i. We consider the case where n DMUs can be divided into two distinct groups with n_l observations ($n = n_1 + n_2$). We assume the following stochastic frontier model for two groups of DMUs:

$$
\begin{aligned}
\text{(The first group)} \quad Y_i &= \phi(\mathbf{X}_i) + \varepsilon_{1,i}, \ i \in \{1, \ldots, n_1\}; \\
\text{(The second group)} \quad Y_i &= \phi(\mathbf{X}_i) + \varepsilon_{2,i}, \ i \in \{n_1 + 1, \ldots, n_1 + n_2\},
\end{aligned}
\quad (1)
$$

where $\varepsilon_{l,i} = V_{l,i} - U_{l,i}$, $V_{l,i}$ is a random noise term of the lth group with $E(V_{l,i}|\mathbf{X}_i) = 0$, and $U_{l,i}$ is an inefficiency term of the lth group with $U_{l,i} \geq 0$ for $l = 1, 2$. We assume that the same production technology is applied to both DMU groups. Hence, the production frontier function $\phi(\cdot)$ is the same throughout the groups, as in Banker et al. [9]. Under this model, we need to estimate the difference $E(U_1) - E(U_2)$ and test the hypothesis

$$
H_0 : E(U_1) - E(U_2) = 0 \text{ vs. } H_1 : E(U_1) - E(U_2) > 0 \, (< 0) \quad (2)
$$

to know which DMU group is more efficient. A novelty of our approach in developing the test is to implement the test without imposing any parametric assumption on the frontier function $\phi(\cdot)$, and with minimal assumptions on inefficiency and random noise. Banker et al. [9] also implemented the test without any parametric assumption on $\phi(\cdot)$, but with additional restrictive parametric assumptions on noise and inefficiency. In the following sections, we first review the work of Banker et al. [9] and then explain the development of our semiparametric regression-based test.

2.1. The Previous Work

To apply the DEA methods to SFM, Banker et al. [9] assumed that the random noise variables $V_{1,i}, V_{2,i}$ are bounded above by V^{max}, that is, $V_{1,i}, V_{2,i} \leq V^{max}$. Under this assumption, they transformed model (1) as

$$
Y_i = (\phi(\mathbf{X}_i) + V^{max}) - (V^{max} - V_{l,i} + U_{l,i}) \equiv \tilde{\phi}(\mathbf{X}_i) - \tilde{U}_{l,i}, \ l = 1, 2. \quad (3)
$$

Since $\tilde{U}_{l,i} = (V^{max} - V_{1,i}) + U_{l,i} \geq 0$, they considered the translated production function $\tilde{\phi}(\cdot) = \phi(\cdot) + V^{max}$ as a new production function, and $\tilde{U}_{l,i}$ as the inefficiency of the DEA framework. The new inefficiency $\tilde{U}_i^l$ can be estimated as $\hat{\tilde{\phi}}(\mathbf{X}_i) - Y_i$ after $\tilde{\phi}(\cdot)$ is estimated using the conventional DEA methods. After estimating $\tilde{U}_{l,i}$ using DEA methods, they used it for group efficiency comparison. This approach is advantageous in that we use the strength of the existing well-developed DEA techniques. However, the approach has one disadvantage in that the tests developed are based not on inefficiency ($U_{l,i}$) itself, but on the inefficiency contaminated by the positive measurement error ($V^{max} - V_{l,i}$) due to random noise. Additionally, the distributional property of the inefficiency estimated using DEA methods is generally hard to derive or quite complicated, making it very difficult to develop a statistical test theory based on estimated inefficiency (estimate of $\tilde{U}_{l,i}$). Hence, we are motivated to develop a test for (2) directly based on inefficiency $U_{l,i}$. We will explain this in the following section.

2.2. The Proposed Test

This section introduces our approach to testing the hypothesis in (2). Unlike Banker et al. [9], we do not require that neither the noise variance nor inefficiency should be equal across groups. Moreover, we allow for distributional difference in the composite error ε and input vector $\mathbf{X}_i$ from the production environmental factors of each group. Specifically, the variance of V_l and mean of inefficiency U_l can differ by the group as well as conditional distribution of $\mathbf{X}_i$, given group l.

First, model (1) can be written as two nonparametric mean regression models as follows:

$$Y_i = [\phi(X_i) - E(U_1)] + [V_{1,i} - (U_{1,i} - E(U_1))]$$
$$\equiv \phi^*(X_i) + \varepsilon^*_{1,i}, \ i \in \{1,\ldots,n_1\}; \quad (4)$$
$$Y_i = [E(U_1) - E(U_2)] + [\phi(X_i) - E(U_1)] + [V_{2,i} - (U_{2,i} - E(U_2))]$$
$$\equiv \beta_0 + \phi^*(X_i) + \varepsilon^*_{2,i}, \ i \in \{n_1+1,\ldots,n_1+n_2\}, \quad (5)$$

where $E(\varepsilon^*_{1,i}) = E(\varepsilon^*_{2,i}) = 0$, and $\beta_0 = E(U_1) - E(U_2)$. If a dummy variable is defined for groups letting $T_i = 0$ for $i \in \{1,\ldots,n_1\}$ and $T_i = 1$ for $i \in \{n_1+1,\ldots,n_1+n_2\}$, the two models (4) and (5) can be integrated into a single partial linear semiparametric regression model as follows:

$$Y_i = \beta_0 T_i + \phi^*(X_i) + \varepsilon^*_i, \ i \in \{1,\ldots,n\}, \quad (6)$$

where $\varepsilon^*_i = (1 - I(T_i = 1))\varepsilon^*_{1,i} + I(T_i = 1)\varepsilon^*_{2,i}$ and $E(\varepsilon^*_i | T_i, X_i) = 0$. Using this model (6), we can test hypothesis (2) by testing hypothesis

$$H_0: \beta_0 = 0 \text{ vs. } H_1: \beta_0 > 0(<0). \quad (7)$$

Note that $Var(\varepsilon^*_i | T_i, X_i) = (1 - I(T_i = 1))Var(\varepsilon^*_{1,i}) + I(T_i = 1)Var(\varepsilon^*_{2,i})$. Thus, model (6) is a heteroscedastic partial linear model. Liang [12] and Ma et al. [13] studied model (6) when X_i is univariate. By extending the theory from there to the case where X_i is multivariate, we can test hypothesis (7). In Appendix A, we prove the asymptotic normality of the kernel-based profile estimator of β_0 based on a local linear model smoother when X_i is multivariate, and provide the necessary assumptions for it. As with the estimator in Liang [12], the kernel-based profile estimator of β_0 when X_i is multivariate is given as

$$\hat{\beta}_0 = (T^\top (I-S)^\top (I-S)T)^{-1} T^\top (I-S)^\top (I-S) Y \equiv HY, \quad (8)$$

where $T = (T_1,\ldots,T_n)^\top$, $Y = (Y_1,\ldots,Y_n)^\top$, and S the smoother matrix for estimating the vector $(E(\cdot|X_1),\ldots,E(\cdot|X_n))^\top$. If we choose local linear regression as the smoothing method, the smoothing matrix $S = [s_{X_1} \cdots s_{X_n}]^\top$ will be a collection of row vectors, each of which is the smoother vector

$$s_x^\top = e_1^\top (X_x^\top W_x X_x)^{-1} X_x^\top W_x, \quad (9)$$

where $e_1^\top = (1,0,\ldots,0)$ is a $(p+1) \times 1$ vector; $W_x = \text{diag}\{K_h(X_1 - x),\cdots,K_h(X_n - x)\}$ for some kernel function K and bandwidth vector $h = (h_1,\ldots,h_p)^\top$; and

$$X_x = \begin{bmatrix} 1 & (X_1 - x)^\top \\ \vdots & \vdots \\ 1 & (X_n - x)^\top \end{bmatrix}. \quad (10)$$

Here, $K_h(X_i - x) = \prod_{j=1}^p h_j^{-1} K((X_{j,i} - x_j)/h_j)$. From the theorem in Appendix A, under some regularity conditions, $\sqrt{n}(\hat{\beta}_0 - \beta_0)$ is asymptotically normal with mean zero and variance $\sigma^2 = E(\tilde{T}^2)^{-2} E(\varepsilon^* \tilde{T})^2$, where $\tilde{T} = T - E(T|X)$. Using a consistent estimator of σ^2, we can test (7) with significance level α by rejecting H_0 if $Z = \hat{\beta}_0/(\hat{\sigma}/\sqrt{n}) \geq z_\alpha$ (or $\leq -z_\alpha$), where z_α is the $(1-\alpha)$-quantile of the standard normal distribution.

As regards the estimation of σ^2, we can first directly estimate variance σ^2 using the estimates $\hat{\varepsilon}^*_i$ and $\hat{E}(T|X_i)$, where $\hat{\varepsilon}^*_i = Y_i - \hat{\beta}_0 T_i - \hat{\phi}^*(X_i)$ and $\hat{\phi}^*(\cdot)$ is the local linear estimator of $\phi^*(\cdot)$ based on $Y_i - \hat{\beta}_0 T_i$, $i = 1,\ldots n$. We can also estimate it using the sandwich covariance estimate based on (8),

$$\widehat{Var}(\hat{\beta}_0 | T, X_1,\ldots, X_n) = H \widehat{Var}(Y|T,X_1,\ldots,X_n) H^\top. \quad (11)$$

Matrix $\widehat{Var}(Y|T, X_1, \ldots, X_n)$ is diagonal, with the ith diagonal element equal to

$$\widehat{E}(\varepsilon^{*2}|T_i, X_i) = (1 - I(T_i = 1))\widehat{Var}(\varepsilon_1^*) + I(T_i = 1)\widehat{Var}(\varepsilon_2^*), \tag{12}$$

where

$$\widehat{Var}(\varepsilon_1^*) = n_1^{-1} \sum_{i:\, T_i=0} (\hat{\varepsilon}_i^*)^2 - \left(n_1^{-1} \sum_{i:\, T_i=0} \hat{\varepsilon}_i^*\right)^2, \tag{13}$$

$$\widehat{Var}(\varepsilon_2^*) = n_2^{-1} \sum_{i:\, T_i=1} (\hat{\varepsilon}_i^*)^2 - \left(n_2^{-1} \sum_{i:\, T_i=1} \hat{\varepsilon}_i^*\right)^2. \tag{14}$$

Since the frontier function is generally (coordinatewise) non-decreasing with respect to the input variables, one might consider it necessary to impose such a monotonicity on $\phi^*(\cdot)$. However, from Theorem 2.1 in Huang [14], such imposition will not decrease the asymptotic variance of $\hat{\beta}_0$; that is, it shows no theoretical improvement in performance. We therefore choose to develop the test without the monotonicity assumption for simplicity.

Note that our test directly estimates the mean difference in inefficiency $E(U_1) - E(U_2)$ using the semiparametric regression technique. Thus, the proposed test can work under assumptions that are more general than those in Banker et al. [9]. Additionally, we do not have to assume that noise has a finite upper support bound (V_{max}). However, the tests in Banker et al. [9] need such assumptions because they estimate $\tilde{U}_{l,i} = (V_{max} - V_{l,i}) + U_{l,i}$ and use it as a surrogate estimate of $U_{l,i}$. However, $V_{max} - V_{l,i}$ may hamper the tests and degrade their performance.

3. Numerical Studies

In this section, we compare the performance of our test with those of Banker et al. [9]. We consider single and multiple input cases and use sandwich formulas to estimate the variance in estimators.

3.1. Single Input Case

We first consider a single input case using the following model:

$$
\begin{aligned}
Y_i &= \phi(X_i) + V_{1,i} - U_{1,i},\ i \in \{1, \ldots, n_1\} \\
Y_i &= \phi(X_i) + V_{2,i} - U_{2,i},\ i \in \{n_1 + 1, \ldots, n_1 + n_2\},
\end{aligned}
$$

where $\phi(x) = 30x - 9x^2$, $X \sim U(0,1)$, $U_{l,i} \sim N_+(0, \sigma_{l,u}^2)$, and $V_{l,i}$ follow the truncated normal distribution with mean 0 and variance $\sigma_{l,v}^2$, which lies within $(-6\sigma_{l,v}, 6\sigma_{l,v})$, $l = 1, 2$. Here, N_+ stands for a normal distribution limited to the domain $[0, \infty)$. As for $V_{l,i}$, we try two cases to reflect both the equal and unequal error variances between groups. We set $\sigma_{1,v} = \sigma_{2,v} = 1$ for the equal error variance case and $\sigma_{1,v} = \sqrt{2}$, $\sigma_{2,v} = 1$ for the unequal variance case. To evaluate the type I error rate and power, we again consider two cases based on whether a mean difference (β_0) exists or does not exist between group inefficiencies: $\sigma_{1,u} = \sigma_{2,u} = 1$ ($\beta_0 = 0$) and $\sigma_{1,u} = \sqrt{2}$, $\sigma_{2,u} = 1$ ($\beta_0 = 0.3305$). Here, the type I error rate implies the rate of supporting group difference in mean inefficiency when there is no difference and the power means the rate of supporting group difference in mean inefficiency when there are really inefficiency differences between groups.

We consider three sample sizes, $n = 100$, 200, and 400; the proportion of each group is approximately 50% and number of replications is 1000. For a comparison, we report the type I error and power of the following five tests with significance level $\alpha = 0.05$: our proposed test (PT), the OLS test, the T-test, the Mann-Whitney (MW) test, the Kolmogorov-Smirnov (KS) test, and the F-test. The last five tests are from Banker et al. [9]. We used a plug-in principle (see Ruppert et al. [15]) to find the bandwidth for our PT.

The test results are depicted in Table 1. Four of these tests, that is, except the KS test and the F-test, seem to respect the significance level in both the equal and unequal variance cases. However, the KS test obviously shows a larger type I error rate than expected for unequal error variances and the F-test seems to be a conservative test, which gives much smaller type I error probabilities than expected. As regards the power, our PT performs best, with the largest power among all the tests. In unequal variance cases where $n = 200, 400$, the KS test has larger power than our PT. However, the KS test is not reliable since it tends to reject the null hypothesis too easily in those cases. Finally, all tests tend to show higher power with larger sample sizes.

Table 1. Type I error and power of the single input case with equal and unequal error variances.

Variances	n	Type I Error (Rejection Rate When $\beta_0 = 0$)						Power (Rejection Rate When $\beta_0 = 0.3305$)					
		PT	OLS	T	MW	KS	F	PT	OLS	T	MW	KS	F
Equal	100	0.052	0.050	0.050	0.050	0.037	0.013	0.377	0.336	0.320	0.283	0.200	0.152
($\sigma_{1,v} = \sigma_{2,v}$)	200	0.062	0.058	0.056	0.062	0.047	0.006	0.595	0.555	0.547	0.483	0.397	0.300
	400	0.063	0.064	0.062	0.061	0.052	0.003	0.848	0.822	0.818	0.761	0.665	0.533
Unequal	100	0.047	0.065	0.060	0.050	0.062	0.037	0.337	0.337	0.328	0.276	0.277	0.279
($\sigma_{1,v} \neq \sigma_{2,v}$)	200	0.053	0.064	0.063	0.052	0.099	0.029	0.505	0.468	0.463	0.388	0.521	0.401
	400	0.044	0.051	0.050	0.048	0.179	0.026	0.738	0.722	0.712	0.645	0.836	0.650

3.2. Multiple Input Case

We next consider a multiple input case with $p = 3$. All the components in the simulation, except for the frontier function ϕ, are the same as in the single input case. As regards the frontier function, we consider two scenarios: the production function has an additive form, and the production function does not have an additive form. The additive assumption on the production function is used in Ferrara and Vidoli [16], but it may not be satisfied in some cases. However, in case of multiple covariates, the practical applicability of our proposed method may become worse since it requires multivariate smoothing and therefore suffers from the well-known "curse of dimensionality" problem as the dimension of the covariates becomes higher. In this case, additive modeling can be a meaningful alternative. For this, we try to estimate the difference in group efficiencies and test whether it is zero with an alternative estimating strategy, where we employ a backfitting procedure, which is a well-known estimating approach under the additive assumption. See Appendix B for details of the alternative method. Considering these two scenarios (additive and non-additive production functions), we examine how our PT(n) and its alternative based on the additive assumption, PT(a), behave depending on the validity of the assumption. The model considered here is

$$Y_i = \phi(X_i) + V_{1,i} - U_{1,i}, \ i \in \{1, \ldots, n_1\}$$
$$Y_i = \phi(X_i) + V_{2,i} - U_{2,i}, \ i \in \{n_1 + 1, \ldots, n_1 + n_2\},$$

where $X_{j,i}$, $j = 1, 2, 3$, are generated from $U(0,1)$ independently. For the first scenario, we set $\phi(x) = (30x_1 - 9x_1^2) + (5 + 2\arctan(10(x_2 - 0.5))) + (4\sqrt{x_3})$; this has an additive structure. For the second scenario, we consider $\phi(x) = (4\sqrt{x_1} + 7\sqrt{x_2} + 5\sqrt{x_3} + 8\sqrt{x_1 x_2} + 10\sqrt{x_2 x_3} + 9\sqrt{x_1 x_3})^{1.1}$. Note that both these production functions are concave. For PT(n), we select bandwidths by a generalized cross validation method (see Hastie and Tibshirani [17]), and for PT(a), we adopt a plug-in principle, as in the single input case.

The performances of PT(n) and PT(a) as well as the other five tests are reported in Tables 2 and 3. From Table 2, our PT(n) outperforms the competitors overall, especially in the unequal variances case. Note that its type I error rates do not deviate much from 0.05, which means that the type I error rate is under control as desired; those of other competitors such as the OLS, T and KS tests tend to be a bit smaller than this level in case of equal variances, and considerably larger in case of unequal variances. The MW test seems to respect the level like PT(n) but PT(n) turns to be more powerful than MW. Note that PT(a) shows good results in terms of type I error rate, with comparable power to the MW test. Its power is lower than that of PT(n), but this is natural since the true production function is not additive. The F-test seems to be anticonservative leading to too high of a type I error probability, especially in the unequal variances case. However, in the equal variance case, the F-test shows shows very good performance in terms of type I error rate and power in large samples, as reported in Banker et al. [9]. From Table 3, our proposed two tests outperform their competitors when the additive assumption is true. Note that under the additive assumption, both PT(a) and PT(n) correctly specify the model. From our simulation, PT(a) slightly outperforms PT(n), since their type I error rates are close to 0.05 and the power of PT(a) is larger than that of PT(n). In case of equal variances, the type I error rates of the five competitors tend to be below 0.05, but when the variances are unequal, their power becomes much lower than our proposed tests although overall they show satisfactory type I errors. The only exception is the F-test. It shows the largest power in the unequal variances case but such merit is dimmed by considerably larger type I errors than other tests.

Table 2. Type I error and power of the multiple input case with equal and unequal error variances when the true production function is not additive.

Variances	n	Type I Error (Rejection Rate When $\beta_0 = 0$)						Power (Rejection Rate When $\beta_0 = 0.3305$)						
		PT(a)	PT(n)	OLS	T	MW	KS	F	PT(a)	PT(n)	OLS	T	MW	KS	F
Equal ($\sigma_{1,v} = \sigma_{2,v}$)	100	0.066	0.053	0.044	0.043	0.054	0.040	0.141	0.220	0.294	0.234	0.235	0.170	0.133	0.492
	200	0.059	0.077	0.049	0.050	0.044	0.035	0.113	0.328	0.488	0.473	0.471	0.356	0.268	0.644
	400	0.048	0.056	0.039	0.039	0.037	0.031	0.064	0.509	0.765	0.705	0.706	0.580	0.517	0.815
Unequal ($\sigma_{1,v} \neq \sigma_{2,v}$)	100	0.061	0.063	0.101	0.099	0.054	0.046	0.316	0.199	0.258	0.346	0.340	0.173	0.143	0.649
	200	0.058	0.074	0.139	0.137	0.062	0.053	0.381	0.302	0.433	0.583	0.580	0.347	0.331	0.849
	400	0.048	0.060	0.128	0.130	0.046	0.093	0.385	0.468	0.682	0.793	0.792	0.530	0.650	0.947

Table 3. Type I error and power of the multiple input case with equal and unequal error variances when the true production function is additive.

Variances	n	Type I Error (Rejection Rate When $\beta_0 = 0$)						Power (Rejection Rate When $\beta_0 = 0.3305$)						
		PT(a)	PT(n)	OLS	T	MW	KS	F	PT(a)	PT(n)	OLS	T	MW	KS	F
Equal ($\sigma_{1,v} = \sigma_{2,v}$)	100	0.054	0.072	0.046	0.045	0.041	0.034	0.065	0.364	0.328	0.213	0.208	0.170	0.132	0.279
	200	0.066	0.060	0.063	0.061	0.056	0.045	0.045	0.578	0.551	0.387	0.384	0.318	0.258	0.403
	400	0.054	0.055	0.034	0.034	0.033	0.033	0.018	0.820	0.801	0.572	0.571	0.503	0.419	0.517
Unequal ($\sigma_{1,v} \neq \sigma_{2,v}$)	100	0.057	0.070	0.066	0.062	0.044	0.048	0.144	0.301	0.295	0.236	0.228	0.153	0.121	0.381
	200	0.068	0.060	0.089	0.089	0.054	0.060	0.137	0.501	0.481	0.415	0.414	0.281	0.290	0.566
	400	0.057	0.054	0.058	0.058	0.032	0.056	0.096	0.719	0.695	0.588	0.586	0.446	0.535	0.720

4. Application to PISA 2015 Data

In this section, we applied our PT to PISA 2015 data and test the efficiency difference between male and female student groups at a specific age in terms of learning time (X_i) and achievement (Y_i) in mathematics. The data can be downloaded from http://www.oecd.org/pisa/data/. PISA is a worldwide study to evaluate educational systems by measuring the scholastic performance of 15-year-old school students in mathematics, science, and reading. We considered the regional averages of the students' learning time and achievement in mathematics based on test results of the 2015 version as production data. Out of the 103 regions in the data, two regions, Nova Scotia in Canada and Chile, were excluded from our analysis in view of their outlier characteristics in efficiency analysis.

Usually, international large-scale assessments data include measurement errors at the individual as well as group level. Therefore, we considered the following stochastic frontier model for such data:

$$\text{(Male Students)} \quad Y_i = \phi(X_i) + V_i^{male} - U_i^{male}, \; i \in \{1, \ldots, 101\}$$
$$\text{(Female Students)} \quad Y_i = \phi(X_i) + V_i^{female} - U_i^{female}, \; i \in \{102, \ldots, 202\}.$$

In this model, we assumed that there would be no gender difference in learning ability from a biological point of view and use the same production frontier for both gender group. It means that all the socio-economic characteristics of differentiation between the gender groups were in the random error terms and not introduced in the frontier function.

Table 4 shows summary statistics of each student group data. We applied the six tests in Section 3 to the data and calculated the p-values for the following hypothesis testing:

$$H_0 : E(U^{male}) = E(U^{female}) \quad \text{vs.} \quad H_1 : E(U^{male}) < E(U^{female})$$

Table 4. Summary statistics of our PISA 2015 data.

		min	Q_1	median	mean	Q_3	max
male	X_i	27.89	39.52	41.72	43.10	47.81	56.70
	Y_i	338.5	470.6	499.7	483.9	513.8	565.6
female	X_i	25.23	38.99	41.49	41.98	45.26	56.67
	Y_i	339.0	456.9	487.9	474.8	501.9	565.0

From Samuelsson and Samuelsson [18], it is known that male students are often more involved in mathematics classes than female students. Additionally, since women are more involved in domestic chores than men and for men time is often made free by their families and relatives for the learning activity, male students are likely to be in an environment where they can focus more on studying than female students. Hence, we expected the results of the test to indicate that the effectiveness of male students was greater than that of female students in average.

Table 5 gives the test results. At a significance level of around $\alpha = 0.05$, our PT, the MW test, and the KS test (with p-value slightly higher that $\alpha = 0.05$) supported the hypothesis that on average male students are more efficient in mathematics than female students. However, the OLS test, the T-test and the F-test reported no significant difference in learning efficiency between the two groups. The reason for this could be the somewhat restrictive assumptions for test validity. Thus, the three tests seem to face the risk of unreasonable results if the assumptions are not satisfied in practice, but our PT does not seem to suffer from this problem.

Table 5. P-values of the six tests to detect efficiency difference in groups of male and female students in terms of learning time and achievement in mathematics.

test	PT	OLS	T	MW	KS	F
p-value	0.049	0.150	0.150	0.044	0.057	0.322

5. Discussion and Conclusions

In this study, we developed tests with sound statistical theory for group efficiency comparison under SFM with considerably better performance than the previous tests proposed in numerical simulations. However, there is still room for improvement in our methods.

First, since we perform full nonparametric modeling for the frontier function $\phi(\cdot)$, which can be multivariate, our test might suffer from the "curse of dimensionality" and require high-order kernels for implementation with four or more input variables. In such a situation, we can consider an alternative test with spirit as in our test when the frontier function $\phi(\mathbf{X})$ has an additive structure, that is, $\phi(\mathbf{X}) = \sum_{j=1}^{p} \phi_j(X_j)$, or could be well-approximated by it.

Second, we only deal with one output case, which limits practical applications. Our methods should be extended to cover the case of multi-output production frontiers, which DEA methods cover.

Third, we assume the same production frontier for both group, which is a clear limitation in practice since such situation is not frequently observed. If it is important to assume separate production frontier functions for different groups, one can use the meta-frontier approach. O'Donnell et al. [6] proposed a meta-frontier approach to compare the group technical efficiencies under stochastic frontier framework. The proposed method has the advantage that it can be used without assuming a common frontier function. However, the use of their method sometimes can be restricted by their assumption that the frontier production function is log-linear.

Finally, if one is interested in estimating the mean inefficiency of each group, we refer to Noh and Van Keiligom [19], which is a recent work along that direction.

Author Contributions: Conceptualization, H.N.; methodology, H.N. and S.J.Y.; software, H.N. and S.J.Y.; formal analysis, H.N. and S.J.Y.; investigation, H.N. and S.J.Y.; writing–original draft preparation, H.N. and S.J.Y.; writing–review and editing, H.N. and S.J.Y.. All authors have read and agree to the published version of the manuscript.

Funding: H. Noh was supported by the Basic Science Research Program through the National Research Foundation of Korea funded by the Ministry of Education (NRF-2017R1D1A1A09000804), and S.J.Y. was supported by research funds for newly appointed professors of Jeonbuk National University in 2018.

Conflicts of Interest: The authors declare no conflicts of interest.

Appendix A

In this appendix, we provide details of the asymptotic normality of the proposed estimator $\hat{\beta}_0$ in Section 2.2. For this, we first list the relevant assumptions.

Assumption

1. The kernel function K is symmetric, and Lipschitz continuous in $[-1, 1]$.
2. ϕ is twice partially continuously differentiable.
3. The density functions of X_j ($j \in \{1, \ldots, p\}$) are continuous, and bounded away from zero and infinity on their supports $\mathcal{C}_j$, which are bounded.
4. V_1, V_2, U_1 and U_2 have finite second moments.
5. For $j \in \{1, \ldots, p\}$, h_j are asymptotic to n^{-a} for $a > 0$ such that $n(\prod_{j=1}^{p} h_j)^2 / \log n \to \infty$ and $nh_j^8 \to 0$ as n goes to infinity.

Theorem A1. *Under the above assumptions,*

$$\sqrt{n}(\hat{\beta}_0 - \beta_0) \longrightarrow N(0, \sigma^2)$$

where

$$\sigma^2 = E(\tilde{T}^2)^{-2} E(\varepsilon^{*2} \tilde{T}^2)$$
$$\tilde{T} = T - E(T|X)$$

Proof. We write

$$\sqrt{n}(\hat{\beta}_0 - \beta_0) = \left[n^{-1} \mathbf{T}^\top (\mathbf{I} - \mathbf{S})^\top (\mathbf{I} - \mathbf{S}) \mathbf{T}\right]^{-1}$$
$$\times n^{-1/2} \left[\mathbf{T}^\top (\mathbf{I} - \mathbf{S})^\top (\varepsilon^* + \boldsymbol{\phi}^* - \mathbf{S}(\mathbf{Y} - \beta_0 \mathbf{T}))\right]$$

where $\boldsymbol{\phi}^* = (\phi^*(X_1), \ldots, \phi^*(X_n))^\top$ and $\varepsilon^* = (\varepsilon_1^*, \ldots, \varepsilon_n^*)^\top$. It suffices to show that

$$n^{-1} \mathbf{T}^\top (\mathbf{I} - \mathbf{S})^\top (\mathbf{I} - \mathbf{S}) \mathbf{T} \xrightarrow{p} E(\tilde{T}^2), \text{ and} \quad (A1)$$

$$n^{-1/2} \left[\mathbf{T}^\top (\mathbf{I} - \mathbf{S})^\top (\varepsilon + \boldsymbol{\phi}^* - \mathbf{S}(\mathbf{Y} - \beta_0 \mathbf{T}))\right] \xrightarrow{d} N(0, E(\epsilon^*) \tilde{T}^2). \quad (A2)$$

To prove these, we first give the following fact.

$$\sup_{\mathbf{x} \in \mathcal{C}_1 \times \cdots \times \mathcal{C}_p} |\hat{\zeta}(\mathbf{x}) - \zeta(\mathbf{x})| = O_p(n^{-2a} + n^{-(1-ap)/2} \log n), \quad (A3)$$

where $\zeta(\mathbf{x}) = E(R|\mathbf{X} = \mathbf{x})$ and $\hat{\zeta}(\mathbf{x})$ is its local linear estimator. That is, $\hat{\zeta}(\mathbf{x}) = s_x^\top \mathbf{R}$ with $\mathbf{R} = (R_1, \ldots, R_n)^\top$ when R_j is a response variable. (A3) can be shown from the standard theory of kernel smoothing. Note that

$$(\mathbf{I} - \mathbf{S}) \mathbf{T} = (\mathbf{T} - E(\mathbf{T}|\mathbf{X})) + (E(\mathbf{T}|\mathbf{X}) - \mathbf{S} \mathbf{T}).$$

The second term of the right-hand side of the above equation is $o_p(1)$ from (A3). This proves (A1). Next, we write

$$n^{-1/2} \left[\mathbf{T}^\top (\mathbf{I} - \mathbf{S})^\top (\varepsilon^* + \boldsymbol{\phi}^* - \mathbf{S}(\mathbf{Y} - \beta_0 \mathbf{T}))\right]$$
$$= n^{-1/2} (\mathbf{T} - E(\mathbf{T}|\mathbf{X}))^\top \varepsilon^* + n^{-1/2} (\mathbf{T} - E(\mathbf{T}|\mathbf{X}))^\top (\boldsymbol{\phi}^* - \mathbf{S}(\mathbf{Y} - \beta_0 \mathbf{T}))$$
$$+ n^{-1/2} (E(\mathbf{T}|\mathbf{X}) - \mathbf{S} \mathbf{T})^\top \varepsilon^* + n^{-1/2} (E(\mathbf{T}|\mathbf{X}) - \mathbf{S} \mathbf{T})^\top (\boldsymbol{\phi}^* - \mathbf{S}(\mathbf{Y} - \beta_0 \mathbf{T}))$$
$$\equiv n^{-1/2} (\mathbf{T} - E(\mathbf{T}|\mathbf{X}))^\top \varepsilon^* + A_{1,n} + A_{2,n} + A_{3,n}.$$

Since $n^{-1/2} (\mathbf{T} - E(\mathbf{T}|\mathbf{X}))^\top \varepsilon^* = N(0, E(\epsilon^*) \tilde{T}^2) + o_p(1)$, it is enough to show that $A_{j,n} = o_p(1)$, $j = 1, 2, 3$, to claim (A2).
To treat $A_{1,n}$, we note that

$$\sup_{x_j \in \mathcal{C}_j} \left| \frac{\partial}{\partial x_j} \hat{\zeta}(\mathbf{x}) - \frac{\partial}{\partial x_j} \zeta(\mathbf{x}) \right| = O_p(n^{-2a} + n^{-(1-ap-2a)/2} \log n), \ j \in \{1, \ldots, p\} \quad (A4)$$

and denote $\hat{\boldsymbol{\phi}}^* = \mathbf{S}(\mathbf{Y} - \beta_0 \mathbf{T})$. Then,

$$A_{1,n} = n^{-1/2} \sum_{i=1}^n (T_i - E(T_i|X_i))(\phi^*(X_i) - \hat{\phi}^*(X_i)).$$

Let $\mathcal{G}$ denote a class of functions satisfying $|g(\mathbf{x}) - g(\mathbf{y})| \leq \|\mathbf{x} - \mathbf{y}\|$ for $\mathbf{x}, \mathbf{y} \in [0,1]^p$. Then, $n^{a_0}(\hat{\phi}^*(\cdot) - \phi^*(\cdot))$ belongs to $\mathbf{G}$ with probability tending to 1 from (A3) and (A4), where $a_0 < \max\{2a, (1 - ap - 2a)/2\}$. We can show that the δ-entropy of $\mathbf{G}$ for the supremum norm satisfies

$$H_\infty(\delta, \mathcal{G}) \leq K\left(\log\frac{1}{\delta} + \frac{1}{\delta^p}\right)$$

for some constant K. Here, we consider an empirical process

$$n^{-1/2}\sum_{i=1}^n (T_i - E(T_i|\mathbf{X}_i))g(\mathbf{X}_i), \ g \in \mathcal{G}$$

indexed by $\mathcal{G}$. Then, $E[(T_i - E(T_i|\mathbf{X}_i))g(\mathbf{X}_i)] = 0$, and by the Corollary 8.8 of van de Geer [20], we conclude that $\sup_{g \in \mathcal{G}}\left|n^{-1/2}\sum_{i=1}^n(T_i - E(T_i|\mathbf{X}_i))g(\mathbf{X}_i)\right| = O_p(1)$, to result in $\mathbf{A}_{1,n} = o_p(1)$. Note that the exponential tail condition, required to apply the empirical process technique, is automatically satisfied in our case since T is a binary variable.

As for $\mathbf{A}_{2,n}$, we first note that $E(\mathbf{A}_{2,n}|(T_1, \mathbf{X}_1), \ldots, (T_n, \mathbf{X}_n)) = 0$. Moreover,

$$\begin{aligned}
&E(\mathbf{A}_{2,n}^2|(T_1, \mathbf{X}_1), \ldots, (T_n, \mathbf{X}_n)) \\
&\leq n^{-1}[var(\varepsilon_1^{1,*}) + var(\varepsilon_1^{2,*})](E(\mathbf{T}|\mathbf{X}) - \mathbf{ST})^\top(E(\mathbf{T}|\mathbf{X}) - \mathbf{ST}) \\
&= O_p(n^{-4a} + n^{-(1-ap)}\log n)
\end{aligned}$$

from (A3). This establishes $\mathbf{A}_{2,n} = o_p(1)$. Finally, $\mathbf{A}_{3,n} = O_p(n^{1/2-4a} + n^{-1/2+ap/2}\log n) = o_p(1)$ from (A3), to complete the proof. □

Appendix B

In this appendix, we describe an alternative test for (2) when the frontier function $\phi(\mathbf{X})$ has an additive structure, that is, $\phi(\mathbf{X}) = \sum_{j=1}^p \phi_j(X_j)$. Under the additive structure assumption, model (1) can be written as two nonparametric mean regression models:

$$\begin{aligned}
Y_i &= \left[\sum_{j=1}^p E\phi_j(X_{j,i}) - E(U_1)\right] + \left[\sum_{j=1}^p (\phi_j(X_{j,i}) - E\phi_j(X_j))\right] + [V_{1,i} - (U_{1,i} - E(U_1))] \\
&\equiv \mu + \sum_{j=1}^p \phi_j^*(X_{j,i}) + \varepsilon_{1,i}^*, \ i \in \{1, \ldots, n_1\};
\end{aligned} \quad (A5)$$

$$\begin{aligned}
Y_i &= \left[\sum_{j=1}^p E\phi_j(X_{j,i}) - E(U_1)\right] + [E(U_1) - E(U_2)] + \left[\sum_{j=1}^p (\phi_j(X_{j,i}) - E\phi_j(X_j))\right] \\
&\quad + [V_{2,i} - (U_{2,i} - E(U_2))] \\
&\equiv \mu + \beta_0 + \sum_{j=1}^p \phi_j^*(X_{j,i}) + \varepsilon_{2,i}^*, \ i \in \{n_1 + 1, \ldots, n_1 + n_2\},
\end{aligned} \quad (A6)$$

where $\phi_j^*(X_j) = \phi_j(X_j) - E\phi_j(X_j)$ and $E(\phi^*(X_j)) = 0$ for $j \in \{1, \ldots, p\}$. If we introduce the same dummy variable T_i as in the single input case, models (A5) and (A6) can be integrated into one single semiparametric regression model, which would be a (heteroscedastic) partial linear additive model:

$$Y_i = \mu + \beta_0 T_i + \sum_{j=1}^p \phi_j^*(X_{j,i}) + \varepsilon_i^*, \ i \in \{1, \ldots, n\}, \quad (A7)$$

where $\varepsilon_i^* = (1 - I(T_i = 1))\varepsilon_{1,i}^* + I(T_i = 1)\varepsilon_{2,i}^*$ and $E(\varepsilon_i^*|T_i, \mathbf{X}_i) = 0$. Partial linear additive models have been studied by several authors; for example, Fan et al. [21], Fan and Li [22], Li [23], and Wei and Liu [24]. For the test, we use the profile least square estimator of β_0 in Wei and Liu [24]. However, Wei and Liu [24] only showed the asymptotic distribution of the estimator of the parametric component vector (in our case, the estimator of $\beta = (\mu, \beta_0)^\top$) under the homoscedasticity assumption of the error (Theorem 2.1 of their paper), and so we extended their result to the heteroscedasticity case.

To introduce the profile least square estimator of β_0 using the method of Wei and Liu [24], we define some notations. Let

$$\mathbf{X}_{des} = \begin{bmatrix} 1 & \mathbf{X}_1 \\ \vdots & \vdots \\ 1 & \mathbf{X}_n \end{bmatrix} = [\mathbf{1}_n, \mathbf{X}], \mathbf{S}_T = \begin{bmatrix} \mathbf{I}_n & \mathbf{S}_1^* & \cdots & \mathbf{S}_1^* \\ \mathbf{S}_2^* & \mathbf{I}_n & \cdots & \mathbf{S}_2^* \\ \vdots & \vdots & \ddots & \vdots \\ \mathbf{S}_p^* & \mathbf{S}_1^* & \cdots & \mathbf{I}_n \end{bmatrix}, \mathbf{C} = \begin{bmatrix} \mathbf{S}_1^* \\ \mathbf{S}_2^* \\ \vdots \\ \mathbf{S}_p^* \end{bmatrix}, \tag{A8}$$

where $\mathbf{S}_k$ is the smoothing matrix for local linear regression with respect to the jth ($j \in \{1,\ldots,p\}$) covariate vector $\mathbf{X}_j = (X_{j,1},\ldots,X_{j,n})^\top$ with kernel function $K(\cdot)$ and bandwidth h_j, $\mathbf{S}_j^* = (\mathbf{I}_n - \mathbf{1}_n\mathbf{1}_n^\top/n)\mathbf{S}_j$, and $\mathbf{1}_n = (1,\ldots,1)^\top$ with length n. Additionally, we define the additive smoother matrix $\mathbf{W}_j$ as $\mathbf{W}_j = \mathbf{E}_j\mathbf{S}_T^{-1}\mathbf{C}$, where $\mathbf{E}_j$ is a partitioned matrix of dimension $n \times np$ with $n \times n$ identity matrix as the jth "block" and zeros elsewhere. Then, the profile least squares estimator of $\beta = (\mu, \beta_0)^\top$ is obtained as the estimator of the coefficient vector β of a synthetic linear regression model

$$Y_i - \tilde{Y}_i = (\mathbf{T}_{des,i} - \tilde{\mathbf{T}}_{des,i})^\top \beta + \varepsilon_i, \tag{A9}$$

where $\mathbf{W}_M = \sum_{j=1}^p \mathbf{W}_j$, $\tilde{\mathbf{Y}} = (\tilde{Y}_1,\ldots,\tilde{Y}_n)^\top = \mathbf{W}_M\mathbf{Y}$ and $\tilde{\mathbf{T}}_{des} = (\tilde{\mathbf{T}}_{des,1},\ldots,\tilde{\mathbf{T}}_{des,n})^\top = \mathbf{W}_M\mathbf{T}_{des}$. Additionally, since $\mathbf{W}_M\mathbf{1}_n = (0,\ldots,0)^\top$, we know that the linear model (A9) becomes

$$Y_i - \tilde{Y}_i = \mu + (T_i - \tilde{T}_i)\beta_0 + \varepsilon_i^*, \tag{A10}$$

where $\tilde{\mathbf{T}} = (\tilde{T}_1,\ldots,\tilde{T}_n)^\top = \mathbf{W}_M\mathbf{T}$. Hence, after a standard calculation in linear model theory, we obtain the profile least squares estimator of β_0 as

$$\hat{\beta}_0 = \left[\mathbf{T}^\top(\mathbf{I}_n - \mathbf{W}_M)^\top(\mathbf{I}_n - \mathbf{J})(\mathbf{I}_n - \mathbf{W}_M)\mathbf{T}\right]^{-1} \mathbf{T}^\top(\mathbf{I}_n - \mathbf{W}_M)^\top(\mathbf{I}_n - \mathbf{J})(\mathbf{I}_n - \mathbf{W}_M)\mathbf{Y}, \tag{A11}$$

where $\mathbf{J} = \mathbf{1}_n\mathbf{1}_n^\top/n$. Using the results to prove Theorem 2.1 in Wei and Liu [24], we show below that under some regularity conditions, estimator $\hat{\beta}_0$ is asymptotically normal with mean zero and variance

$$\sigma_{add}^2 = E(\tilde{T}^2)^{-2} E(\varepsilon^{*2}\tilde{T}^2), \tag{A12}$$

where $\tilde{T} = T - E(T) - \sum_{j=1}^p \left[E(T|X_j) - E(T)\right]$. Once we have a consistent estimate of σ_{add}^2, we can test (7) for a given significance level α. As with the case of single input variable, we can directly estimate the variance $\sigma_{\beta_0}^2$ using estimates $\hat{\varepsilon}_i^*$ and $\hat{E}(T|X_{j,i})$. Here, $\hat{E}(T|X_{j,i})$ can be obtained as the ith element of $\mathbf{S}_j(T_1,\ldots,T_n)^\top$. Alternatively, we can estimate the variance via the sandwich formula estimate based on (A11) following similar steps in Section 2.2.

Now, we can show the asymptotic property of the profile least square estimator of β_0. For this, we first list the relevant assumptions.

Assumption

1. The kernel function K is symmetric, and Lipschitz continuous in $[-1,1]$.
2. ϕ_j ($j \in \{1,\ldots,p\}$) are twice continuously differentiable.
3. The density functions of X_j ($j \in \{1,\ldots,p\}$) are continuous, and bounded away from zero and infinity on their supports, which are bounded.
4. V_1, V_2, U_1 and U_2 have finite second moments.
5. For $j \in \{1,\ldots,p\}$, $h_j \to 0$, $nh_j/\log n \to \infty$ and $nh_j^8 \to 0$ as n goes to infinity.

Theorem A2. *Under the above assumptions,*

$$\sqrt{n}(\hat{\beta}_0 - \beta_0) \longrightarrow N(0, \sigma_{add}^2)$$

where

$$\sigma_{add}^2 = E(\tilde{T}^2)^{-2} E(\varepsilon^{*2} \tilde{X}^2)$$

$$\tilde{T} = T - E(T) - \sum_{j=1}^{p} [E(T|X_j) - E(T)]$$

Proof. $\hat{\beta}_0$ can be expressed as follows:

$$\hat{\beta}_0 = \left[\mathbf{T}^\top (\mathbf{I}_n - \mathbf{W_M})^\top (\mathbf{I}_n - \mathbf{J})(\mathbf{I}_n - \mathbf{W_M})\mathbf{T}\right]^{-1} \mathbf{T}^\top (\mathbf{I}_n - \mathbf{W_M})^\top (\mathbf{I}_n - \mathbf{J})(\mathbf{I}_n - \mathbf{W_M})\mathbf{Y}$$

where $\mathbf{T} = (T_1, \ldots, T_n)^\top$ and $\mathbf{J} = \mathbf{1}_n \mathbf{1}_n^\top / n$. Then,

$$\begin{aligned}
&\sqrt{n}(\hat{\beta}_0 - \beta_0) \\
&= \left[n^{-1} \mathbf{T}^\top (\mathbf{I}_n - \mathbf{W_M})^\top (\mathbf{I}_n - \mathbf{J})(\mathbf{I}_n - \mathbf{W_M})\mathbf{T}\right]^{-1} \\
&\quad \times n^{-1/2} \mathbf{T}^\top (\mathbf{I}_n - \mathbf{W_M})^\top (\mathbf{I}_n - \mathbf{J})(\mathbf{I}_n - \mathbf{W_M})(\boldsymbol{\phi}^* + \boldsymbol{\varepsilon}^*),
\end{aligned}$$

where $\boldsymbol{\phi}^* = (\sum_{j=1}^{p} \phi^*(X_{j,1}), \ldots, \sum_{j=1}^{p} \phi^*(X_{j,n}))^\top$ and $\boldsymbol{\varepsilon}^* = (\varepsilon_1^*, \ldots, \varepsilon_n^*)^\top$. Here, the term associated with the intercept μ vanishes because $\mathbf{W_M} \mathbf{1}_n = (0, \ldots, 0)^\top$. To prove the theorem, it suffices to show that

$$n^{-1} \mathbf{T}^\top (\mathbf{I}_n - \mathbf{W_M})^\top (\mathbf{I}_n - \mathbf{J})(\mathbf{I}_n - \mathbf{W_M})\mathbf{T}$$
$$= n^{-1} \sum_{i=1}^{n} \left[T_i - E(T_i) - \sum_{j=1}^{p} [E(T_i|X_{j,i}) - E(T_i)]\right]^2 + o_p(1) \quad \text{(A13)}$$

and

$$n^{-1/2} \mathbf{T}^\top (\mathbf{I}_n - \mathbf{W_M})^\top (\mathbf{I}_n - \mathbf{J})(\mathbf{I}_n - \mathbf{W_M})(\boldsymbol{\phi}^* + \boldsymbol{\varepsilon}^*)$$
$$= n^{-1/2} \sum_{i=1}^{n} \left[T_i - E(T_i) - \sum_{j=1}^{p} [E(T_i|X_{j,i}) - E(T_i)]\right] \varepsilon_i^* + o_p(1). \quad \text{(A14)}$$

Note that $(\mathbf{I}_n - \mathbf{J})(\mathbf{I}_n - \mathbf{W_M})\mathbf{X} = (\mathbf{I}_n - \mathbf{W_M})\mathbf{X} - \mathbf{1}_n \bar{X}$, where $\bar{X} = n^{-1} \sum_{i=1}^{n} X_i$ because fact $\mathbf{1}_n^\top \mathbf{W_M} = (0, \ldots, 0)$. Then, one can easily see that

$$n^{-1} \mathbf{X}^\top (\mathbf{I}_n - \mathbf{W_M})^\top (\mathbf{I}_n - \mathbf{J})(\mathbf{I}_n - \mathbf{W_M})\mathbf{T}$$
$$= n^{-1} (\mathbf{T} - \mathbf{1}_n \mu_X)^\top (\mathbf{I}_n - \mathbf{W_M})^\top (\mathbf{I}_n - \mathbf{W_M})(\mathbf{T} - \mathbf{1}_n \mu_T) + O_p(n^{-1/2})$$

for $\mu_T = E(T_1)$. Therefore, Equation (A13) can be verified as in the proof of Lemma 6.2 in Wei and Liu (2012). For Equation (A14), note that

$$n^{-1/2} \mathbf{T}^\top (\mathbf{I}_n - \mathbf{W_M})^\top (\mathbf{I}_n - \mathbf{J})(\mathbf{I}_n - \mathbf{W_M}) \boldsymbol{\phi}^*$$
$$= n^{-1/2} (\mathbf{T} - \mathbf{1}_n \mu_T)^\top (\mathbf{I}_n - \mathbf{W_M})^\top (\mathbf{I}_n - \mathbf{W_M})(\boldsymbol{\phi}^* - \mathbf{1}_n \mu_{\phi^*}) + O_p(n^{-1/2}) \quad \text{(A15)}$$

and

$$n^{-1/2} \mathbf{T}^\top (\mathbf{I}_n - \mathbf{W_M})^\top (\mathbf{I}_n - \mathbf{J})(\mathbf{I}_n - \mathbf{W_M}) \boldsymbol{\varepsilon}^*$$
$$= n^{-1/2} (\mathbf{T} - \mathbf{1}_n \mu_X)^\top (\mathbf{I}_n - \mathbf{W_M})^\top (\mathbf{I}_n - \mathbf{W_M}) \boldsymbol{\varepsilon}^* + O_p(n^{-1/2}), \quad \text{(A16)}$$

where $\mu_{\phi^*} = E(\sum_{j=1}^p \phi^*(X_{j,1}))$. Then, we have

$$(\mathbf{I}_n - \mathbf{W_M})\varepsilon^* = \varepsilon^* - \sum_{j=1}^p \mathbf{S}_j \varepsilon^* + O_p\left(n \sum_{j=1}^p h_j^4\right)$$

from Lemma B.6 in [25]. Note that this is true as long as the conditional variance of the ε^* given covariates exists. This is guaranteed by assumption 4. Wei and Liu (2012) used a similar fact under the homoscedastic error assumption. Then, with a derivation similar to that in Lemma 6.3 of Wei and Liu (2012), we can show that (A15) converges to zero in probability and (A16) can be written as:

$$n^{-1/2} \sum_{i=1}^n \left[T_i - E(T_i) - \sum_{j=1}^p \left[E(T_i|X_{j,i}) - E(T_i) \right] \right] \varepsilon_i^* + o_p(1),$$

to complete the proof. □

References

1. Golany, B.; Storberg, J. A data envelopment analysis of the operational efficiencies of bank branches. *Interfaces* **1999**, *29*, 14–26. [CrossRef]
2. Lee, H.; Park, Y.; Choi, H. Comparative evaluation of performance of national R&D programs with heterogeneous objectives: A DEA approach. *Eur. J. Oper. Res.* **2009**, *196*, 847–855.
3. Cummins, J.D.; Weiss, M.A.; Zi, H. Organizational form and efficiency: The coexistence of stock and mutual property-liability insurers. *Manag. Sci.* **1999**, *45*, 1254–1269. [CrossRef]
4. Simar, L.; Zelenyuk, V. On testing equality of distributions of technical efficiency scores. *Econom. Rev.* **2006**, *25*, 497–522. [CrossRef]
5. Li, Q. Nonparametric testing of closeness between two unknown distribution functions. *Econom. Rev.* **1996**, *15*, 261–274. [CrossRef]
6. O'Donnell, C.J.; Rao, D.S.P.; Battese, G.E. Metafrontier frameworks for the study of firm-level efficiencies and technology ratios. *Empir. Econ.* **2008**, *34*, 231–255. [CrossRef]
7. Aigner, D.; Chu, S. On estimating the industry production function. *Am. Econ. Rev.* **1968**, *58*, 826–839.
8. Meeusen, W.; van den Broeck, J. Efficiency estimation from Cobb-Douglas production functions with composed error. *Int. Econ. Rev.* **1977**, *18*, 435–444. [CrossRef]
9. Banker, R.D.; Zheng, Z.; Natarajan, R. DEA-based hypothesis tests for comparing two groups of decision making units. *Eur. J. Oper. Res.* **2010**, *206*, 231–238. [CrossRef]
10. Simar, L.; Wilson, P.W. Two-stage DEA: Caveat emptor. *J. Product. Anal.* **2011**, *36*, 205–218. [CrossRef]
11. Banker, R.D.; Natarajan, R. Evaluating contextual variables sffecting productivity using data envelopment analysis. *Oper. Res.* **2008**, *56*, 48–58. [CrossRef]
12. Liang, H. Estimation in partially linear models and numerical comparisons. *Comput. Stat. Data Anal.* **2006**, *50*, 675–687. [CrossRef] [PubMed]
13. Ma, Y.; Chiou, J.M.; Wang, N. Efficient semiparametric estimator for heteroscedastic partially linear models. *Biometrika* **2006**, *93*, 75–84. [CrossRef]
14. Huang, J. A note on estimating a partly linear model under monotonicity constraints. *J. Stat. Plan. Inference* **2002**, *107*, 343–351. [CrossRef]
15. Ruppert, D.; Sheather, S.J.; Wand, M.P. An effective bandwidth selector for local least squares regression. *J. Am. Stat. Assoc.* **1995**, *90*, 1257–1270. [CrossRef]
16. Ferrara, G.; Vidoli, F. Semiparametric stochastic frontier models: A generalized additive model approach. *Eur. J. Oper. Res.* **2017**, *258*, 761–777. [CrossRef]
17. Hastie, T.J.; Tibshirani, R.J. *Generalized Additive Models*; Chapman and Hall/CRC: New York, NY, USA, 1990.
18. Samuelsson, M.; Samuelsson, J. Gender differences in boys' and girls' perception of teaching and learning mathematics. *Open Rev. Educ. Res.* **2016**, *3*, 18–34. [CrossRef]
19. Noh, H.; Van Keilegom, I. On relaxing the distributional assumption of stochastic frontier models. *J. Korean Stat. Soc.* **2020**, in press.

20. Van de Geer, S. *Empirical Processes in M-Estimation*; Cambridge University Press: Cambridge, UK, 2000.
21. Fan, J.; Hardle, W.; Mammen, E. Direct estimation of low-dimensional components in additive models. *Ann. Stat.* **1998**, *26*, 943–971. doi:10.1214/aos/1024691083. [CrossRef]
22. Fan, Y.; Li, Q. A kernel-based method for estimating additive partially linear models. *Stat. Sin.* **2003**, *13*, 739–762.
23. Li, Q. Efficient estimation of additive partially linear models. *Int. Econ. Rev.* **2000**, *41*, 1073–1092. [CrossRef]
24. Wei, C.H.; Liu, C. Statistical inference on semi-parametric partial linear additive models. *J. Nonparametric Stat.* **2012**, *24*, 809–823. [CrossRef]
25. Fan, J.; Jiang, J. Nonparametric inferences for additive models. *J. Am. Stat. Assoc.* **2005**, *100*, 890–907. [CrossRef]

© 2020 by the authors. Licensee MDPI, Basel, Switzerland. This article is an open access article distributed under the terms and conditions of the Creative Commons Attribution (CC BY) license (http://creativecommons.org/licenses/by/4.0/).

MDPI
St. Alban-Anlage 66
4052 Basel
Switzerland
Tel. +41 61 683 77 34
Fax +41 61 302 89 18
www.mdpi.com

Mathematics Editorial Office
E-mail: mathematics@mdpi.com
www.mdpi.com/journal/mathematics

www.ingramcontent.com/pod-product-compliance
Lightning Source LLC
LaVergne TN
LVHW071958080526
838202LV00064B/6783